Population Genetics - From DNA to Evolutionary Biology

Edited by Payam Behzadi

Published in London, United Kingdom

Population Genetics - From DNA to Evolutionary Biology
http://dx.doi.org/10.5772/intechopen.107806
Edited by Payam Behzadi

Contributors
A K M Shafiul Kadir, Alekxander Zamotajlov, Alexander Alekseevich Ilyichev, Alexandra Dias, Alireza Tajik, Ana Carvalho, Ashik Sharfaraz, Atoum Abdullah, Ayesha Afrose Ura, Ekaterina Aleksandrovna Volosnikova, Ety Akhter, Evgenyi Khomitskyi, Felicia Ikolo, Fidelina Gonzalez, Francisco Ponce, Imre Lázár, Jessie M. Cameron, Jitendra Kumar, José Lima-Brito, José Luís Lousada, Joye Kundu, Larisa Ivanovna Karpenko, Lavanya Chandramouli, Maria Emília Calvão Moreira Emília Silva, Maria João Gaspar, Mariya Borisovna Borgoyakova, Mary C. Maj, Mohammad Ullah Shemanto, Nitish Krishna Das, Patricio Barria, Pooja SureshKumar, Raisa Sukhodolskaya, Rifgat Shagidullin, Robert Finn, Sabyasachi Maity, Semra Sümer Sertyel, Sergio Mora, Shapaat Kushalieva, Soumik Tripura, Tamara Avtaeva, Ting Zhai, Vidya Niranjan, Zachary D. Nagel

First published in London, United Kingdom, 2024 by IntechOpen
IntechOpen is the global imprint of INTECHOPEN LIMITED, registered in England and Wales, registration number: 11086078, 167-169 Great Portland Street, London, W1W 5PF, United Kingdom

British Library Cataloguing-in-Publication Data
A catalogue record for this book is available from the British Library

Additional hard and PDF copies can be obtained from orders@intechopen.com

Population Genetics - From DNA to Evolutionary Biology
Edited by Payam Behzadi
p. cm.

This title is part of the Genetics Book Series, Volume 1
Topic: Genomics
Series Editor: Kenji Ikehara
Topic Editor: Irina Vlasova-St. Louis

Print ISBN 978-0-85014-057-6
Online ISBN 978-0-85014-058-3
eBook (PDF) ISBN 978-0-85014-059-0
ISSN 3049-7094

For EU product safety concerns:
IN TECH d.o.o., Prolaz Marije Krucifikse Kozulić 3, 51000 Rijeka, Croatia,
info@intechopen.com or visit our website at intechopen.com.

IntechOpen Book Series
Genetics
Volume 1

Aims and Scope of the Series

"Genetics," which has been proud of its tradition since Mendel presented his research results in 1865, initially progressed quite slowly due to simple observational approaches of individuals and groups. However, the discovery of double-stranded DNA by Watson and Crick about 70 years ago triggered rapid progress in life sciences, including genetics, which was primarily conducted using *Escherichia coli* and bacteriophages infecting *E. coli*. Subsequently, genetics has achieved remarkable developments, such as understanding genetic disorders, including cancers, through research on the biogenesis and differentiation of plants and animals.
The two topics of this book series - Human Genetics, and Genomics - will address important areas of advancement in genetics.
Human Genetics: After fundamental genetics, initially studied with the main goal of revealing the functions of individual genes and proteins, genetics expanded from understanding the genetic system itself to understanding many infectious diseases caused by bacteria and viruses. Consequently, human beings are now overcoming infectious diseases by developing medicinal chemicals, including antibiotics and vaccines. However, genetic disorders remain challenging to cure up to now. Nevertheless, even the cure for them, including various cancers, is coming closer to reality due to the rapid progress of human genetics. In this way, the welfare of human life continues to improve, and even longevity, which was once a dream, has been achieved to some extent in recent years.
Genomics: On the other hand, the understanding of the comprehensive interrelationship of whole genes or whole proteins functioning in one organism has become possible now, as research has entered the era of genomics, owing to the rapid progress of base sequence analysis and bioinformatics. The development of genomics has further made it possible to understand the evolutionary processes of organisms through comparative studies among the genomes of many organisms.
This book series will discuss the findings obtained during the advancement of human genetics and genomics. It is also expected that this series will trigger the formation of a better world composed of human beings and all other organisms on Earth through discussions of research results obtained under the development of general genetics.

Meet the Series Editor

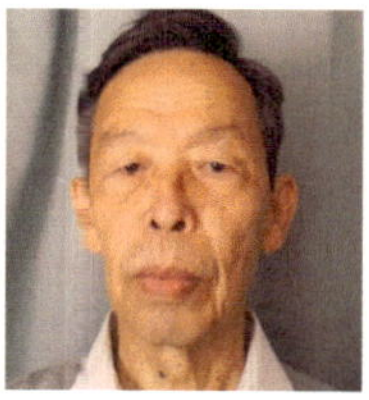

Kenji Ikehara graduated from the Department of Industrial Chemistry, Faculty of Engineering, Kyoto University in 1968. He received his B. Eng. (1968) and subsequently earned M. Eng. (1970) and D. Eng. (1976) degrees from Kyoto University. He began his career as a research associate in the Faculty of Science at the University of Tokyo before moving on to become an associate professor in the Faculty of Science at Nara Women's University. He was later promoted to professor and subsequently served as the dean of the Faculty of Science at Nara Women's University. Additionally, he held the position of director at the Nara Study Center of the Open University of Japan. For approximately 15 years, he focused his research on sporulation initiation of Bacillus subtilis. Later, he shifted his focus to the origins and evolutionary processes of microbial genes, the genetic code, proteins, and life. He has proposed several hypotheses, including the GC-NSF(a) hypothesis on the origin of genes, the GNC-SNS hypothesis on the genetic code, the protein 0th-order structure hypothesis on the origin of proteins, and the [GADV]-protein world hypothesis (GADV hypothesis) on the origin of life. Furthermore, he served as the local chair of the International Conference, Origin 2014, held in Nara in 2014.

Meet the Volume Editor

Dr. Payam Behzadi has been a faculty member at the Department of Microbiology, College of Basic Sciences, Shahr-e-Qods Branch, Islamic Azad University, Iran, since 2004. He has a BSc and MSc in Microbiology and a Ph.D. in Molecular Biology. He is currently an assistant professor at Islamic Azad University. He has authored and edited more than 25 books and book chapters and more than 100 original and review articles. His scientific research interests include urinary tract infections, antibiotics, bioinformatics, genetics, gene profiling, molecular biology, and cellular and molecular immunology. Dr. Behzadi trains as an ice skater in his free time.

Contents

Preface

The genomic pool is an invaluable biological treasure trove associated with evolutionary biology. The genomic structure depicts certain sequences in DNA molecules, which are directly involved in encoding the related information in genetic language (**Figure 1**). In this regard, comparative genomics enables us to focus on molecular bases and evolutionary origins [1]. Furthermore, the study of comparative population genetics supports us in identifying the functional activities of both hypothetical and new genes [2]. The study of genomic elements, including DNA molecules, may provide a deep and rigorous understanding of what happens in population genetics and the different kinds of signatures that can provide accurate perspectives in the field of evolutionary biology across diverse organisms, ranging from simple to complex levels and from unicellular to multicellular entities.

This book is a collection of ten chapters written by international scientists. It is organized into six sections. Section 1 includes Chapter 1, "Coevolutionary Dynamism of Man-Environment-Organism". Section 2 includes Chapter 2, "Recent Advancement on In-Silico Tools for Whole Transcriptome Analysis" and Chapter 3, "Approaches to Improve the Immunogenicity of Plasmid DNA-Based Vaccines against COVID-19" Section 3 includes Chapter 4, "Useful Molecular and Cytogenetic Approaches in

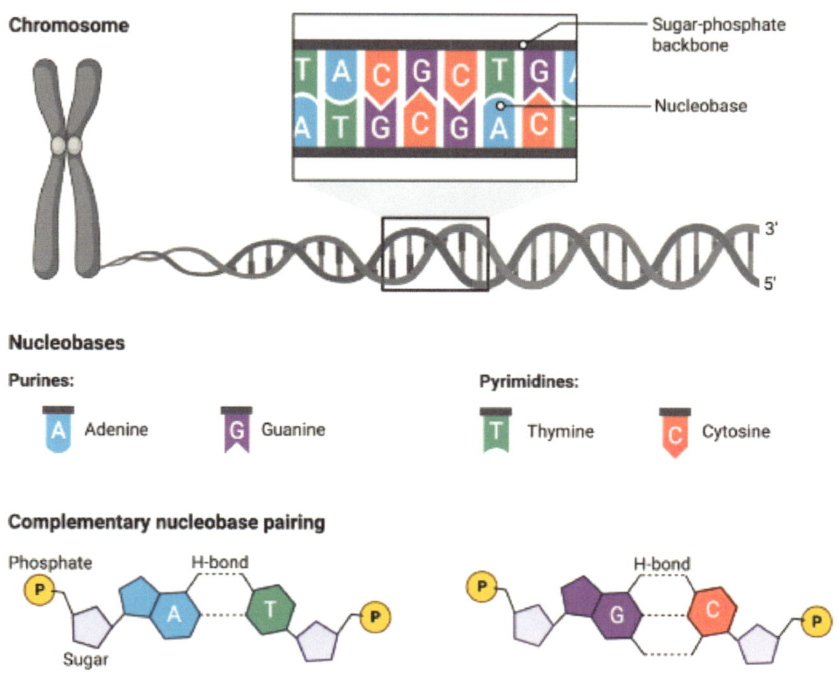

Figure 1.
The structure of a DNA molecule as the basic genetic material in genomics, population genetics, and evolutionary biology (biorender.com).

Population Genetics Studies of Pine Species" and Chapter 5, "Sexual Dimorphism in Physiological Reactions to Biotope Type (the Case Study in Ground Beetles)" Section 4 includes Chapter 6, "Population Genetic Structure of Marine Fishes" and Chapter 7, "Founder Effect: Breeding a Dog for the Elderly Gentleman Reveals an Animal Model of a Human Genetic Disorder" Section 5 includes Chapter 8, "Genetic Revelation of the Potentially Malignant Disorders in the Oral and Maxillofacial Region" Finally, Section 6 includes Chapter 9, "The Hidden Factor of Azoospermia: Telomere Effect" and Chapter 10, "Current Technologies for Measuring or Predicting Telomere Length from Genomic Datasets" As editor, I believe that this book will be a useful academic resource for colleagues worldwide.

I would like to thank the staff at IntechOpen, especially Publishing Process Manager Kristina Kardum Cvitan who supported me throughout this project.

Payam Behzadi
Assistant Professor,
Department of Microbiology,
Shahr-e-Qods Branch,
Islamic Azad University,
Tehran, Iran

References

[1] De la Fuente R, Díaz-Villanueva W, Arnau V, Moya A. Genomic signature in evolutionary biology: A review. Biology. 2023;**12**(2):322

[2] Shoemaker WR, Chen D, Garud NR. Comparative population genetics in the human gut microbiome. Genome Biology and Evolution. 2022;**14**(1):evab116

Section 1

Evolution and Dynamism

Chapter 1

Coevolutionary Dynamism of Man-Environment-Organism

Imre Lázár

Abstract

In our co-evolutionary concept, we reconsider the human-environment unity framed in the M-E-O (Man-Environment-Organism) model, adapting Latour's ANT theory, where the subject of human evolution is seen in unity with its (his/her/their) "Umwelt," creating particular social, memetic, and technospherial environmental extensions and hybrids exposed to mutual selective forces. We analyze this issue in the context of coevolutionary mechanisms influencing genetic and memetic selection. Linguistic samples, the sociocultural aspects of reproduction, or sociocultural answers to the challenge of pandemics, prove the coevolutionary significance of the human ecological approach. The competitive M-E-O complexes are actors and subjects of the selective dynamism of human evolution. The M-E-O model offers a hermeneutic framework to understand the selective evolutionary dynamism of today's techno-civilizational changes, as an accelerated evolutionary process.

Keywords: coevolution, M-E-O model, actor-network theory, neo-evolutionist anthropology, linguistic coevolution, attachment theory, COVID-19 pandemic

1. Introduction

The M-E-O (Man-Environment-Organism) concept [1] as a human ecological model is made up of nests and niches of natural, social, technological, and infospheric environments that are connected and surround the human actor and its communities. The M-E-O model in the present coevolutionary context implies Kenneth Boulding's theory of ecodynamics [2], the General Theory of Evolution by Vilmos Csányi [3], Bruno Latour's Actor-Network Theory and his hybrid concept [4], and coevolutionary processes based on the concept of niche construction [5] in which interacting natural, social, technological, and infospheric niches influence each other in an ecodynamic (competitive, predating, symbiotic, neutral, or saprophytic) manner through human agency. This transactive network of four (natural, social, technological, and infospheric) poles generates the event space to visualize the dynamics of coexistence, and offers a schematic "co-ontology." (**Figure 1**) [7].

The niche construction-based neo-evolutionary framework might help to bridge these different frames of understanding based on the coevolutionary reciprocity of sociocultural changes and evolution. According to Dawkins, organisms can directly transmit environmental artifacts to the next generation with niche constructive

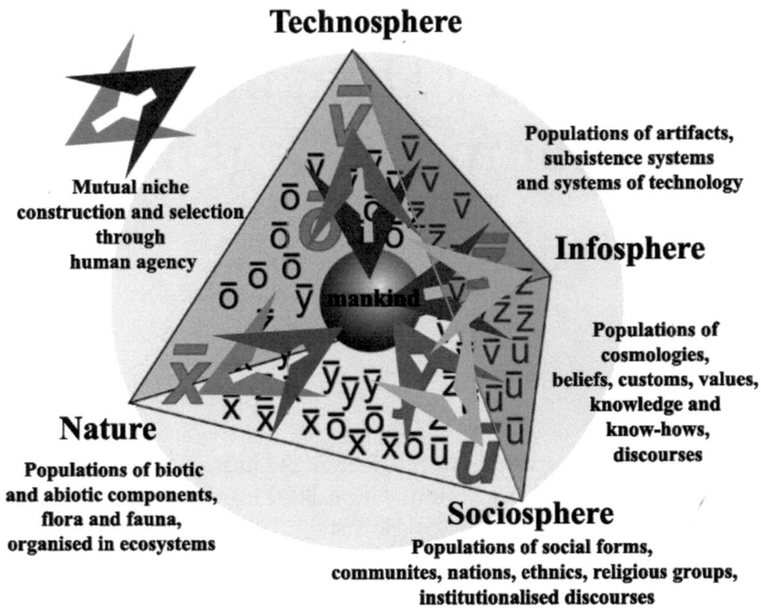

Figure 1.
Mutual niche construction and selection of artifacts, ideas, and social forms through human agency [6].

potential, like the built environment, technological know-how transferred by imitative learning or other ways, and social structures (e.g., mating and hunting practices). *Niche construction* influences epigenetic transmission, including gene-meme interactions and gene–environment coevolution.

The four-dimensional concept of evolution is based on the interrelationships of genetic, epigenetic, behavioral, and symbolic inheritance systems [7]. It implies the tetrahedron M-E-O model with its natural (genetic and epigenetic), behavioral, social and cultural (symbolic inheritance), and technological (influencing all inheritance systems) dimensions.

The M-E-O model (Lázár, 1997) [1, 6] and the ANT (Actor Network Theory) in the 1980s, created by Callon [8], Latour [9], Law [10], and Latour's "hybrid" concept [4], show converging elements, with a common conclusion that modernist nature/culture dualism is no longer an appropriate framework to describe our reality. The growth and structure of knowledge could be analyzed and interpreted through the interactions of actors and networks. Actants denote human and nonhuman actors in a network of relations based on their functional relevance. One might map the relations that are simultaneously material (between things) and infospherical (as concepts in semiotic terms). ANT assumes that many relationships are both material and semiotic. According to Latour, elements of the actor-network also have at least three attributes: "*The first one is to attribute to them naturality, and to link them with nature. The second one is to grant them sociality, and to tie them to the social fabric. The third one is to consider them as a semiotic construction, and to relate agency with building of meaning.*" [10]. The clusters of actors involved in creating meaning are both material and semiotic and usually embedded in the technospherical context of human practice.

2. Methodology

The review of coevolutionary concepts helps to integrate the human ecological approach, the neoevolutionary school of social and cultural anthropology, and the Actor-Network theory in the so-called M-E-O framework. The efforts to integrate archaeogenetic information about the population and memetic aspects of language history; analysis of sociocultural aspects of reproduction with special regard to culture of attachment, and finally a genetic/memetic system's response to catastrophes like COVID-19 in a coevolutionary framework help to draw a common basis for human ecological understanding of biocultural adaptation and selection.

3. Coevolution in biology

The term "coevolution" refers to the mutually reinforcing evolutionary changes brought about by encounters between different species, and can apply to both genetic and memetic replication. The strength of natural selection is proportional to the strength of the relationship between fitness and the value of the trait.

It is possible to keep the discussion of coevolution at the level of biological species and even expand it to include social areas. Coevolutionary insights have a hundred-year history. Kropotkin's theory of evolution [11], based on interspecies cooperation, provides a solid paradigmatic foundation for coevolutionary studies. To investigate coevolution in mutualistic networks, Guimarães et al. [12] integrated data from 20 plant-animal assemblages with a trait evolution model. According to his findings, species-rich mutualisms show three basic characteristics of coevolutionary significance. First, the coevolution of species within mutualistic networks accelerates the rate of evolution, which has an impact on the changing characteristics of these species. Second, coevolution causes traits of species within the same trophic level to converge and interact with one another more often. Thirdly, the presence of super-generalist species, or species that interact with multiple groups of species, is correlated with greater levels of convergence. These conclusions are similar to the conclusions of General Evolution Theory proposed by Csányi and Kampis [13], whose Dynamical Replicative Network Theory identifies a transition from the predominance of parametric information to the expansion of functional information of constituent parts of the interacting coevolutionary compartments.

Paleobiologists use phylogenetic analyses of extant species to probe long-term, macroevolutionary processes of coevolution. Population genetics, quantitative genetics, and optimality theory are all used in computational and mathematical models of coevolutionary processes within populations and species.

Eco-evolutionary research also focuses on more distant forms of coevolution, such as gene-for-gene and diffuse coevolution, in addition to more direct forms of coevolution, such as paired coevolution (or specific coevolution). Diffuse coevolution (sometimes called "guild" coevolution) refers to evolutionary reactions among groups of organisms that are mutually advantageous. This type of coevolution highlights how many species exert complex sets of selective pressures on most species, and how the evolutionary responses of certain species change the conditions under which other species evolve.

Correspondence between genes in different species is an example of gene-for-gene coevolution, sometimes known as "matching gene" coevolution. In his detailed paper,

Domingo [14] talks about the ways that organisms and their pathogens change each other and adapt as a result of their interactions as biological systems that have shared space for a long time. At the macroecological level, we also find coevolutionary rules guiding complex ecosystems, whereas host-parasitic, microbial, or viral interactions open the coevolutionary theater of the microworld.

Thompson and Cunningham [15] say that coevolution, which they define as the mutual adaptation of organisms that interact with each other, shows important processes that link the genomes of interacting species to organize biodiversity and make more diversity through coevolutionary selection, which makes memes and genes coevolve.

A single extended genetic/memetic coevolutionary framework could be used to explain both biological and cultural evolution if we apply these ideas from biological ecosystems to human ecological phenomena. We can employ Dawkins' [16] terminology to discuss memetic evolution. He stated that we "*need a word for the new replicator, a noun that expresses the idea of a unit of cultural transmission or a unit of imitation. Memes can be songs, phrases, catchphrases, trends in clothing, or even techniques for producing pots or constructing arches.*" Memes, according to Blackmore [17], are "*anything that is repeated with variation from one person to another, including behaviors, talents, songs, stories, and other types of information.*" Roland Barthes used the similar term "*narreme*" to describe the narrative of a memetic bundle.

Additionally, this replication process raises the possibility of variation, distortion, and other memetic "mutations." Theories, dogmas, and canonized conceptions may represent a form of climax state for meme complex constituents. Velikovsky [18], similarly to Kuhn's scientific paradigm model, thinks of the periods of consolidation and elaboration as part of the ongoing knowl¬edge evolution, where, according to DiCarlo, the consolidation of memetic patterns, a memetic equilibrium mirroring adapational success, emerges as a result of solving environmental problems, which might lead to rigidity and canonized orthodoxy sometimes [19].

Cultural change, the progress of discourses, and technological inventions are results of mutations of memetic evolution, emerging from repetitions and replications, memetic transfers. Nevertheless, the diverse memetic patterns at the level of personal memetic interactions generate a convergent order in the given culture [20]. Memes (embodied in the technological sector, social discourses and institutions, and art, science, and religion, among other things) emerge as a result of the interaction of three types: copies, variation, and competition for survival. Languages, scientific theories, technological know-hows, laws, religions, books, movies, songs, and other cultural artifacts might be seen as memeplexes, or co-adapted meme complexes, which are replicated and transferred together. According to the information provided, isolated, interbreeding, and local social subpopulations may carry spherical cultural meme patterns. If we borrow a phrase from biology, the so-called "*deme*," they are selected as a group-level unit rather than at the level of an individual. Under intense pressure from the economy, technology, or politics, the memetic patterns (clothing, customs, belief systems, rituals, and religion) of such a local population unit ("*deme*" signifying the population of a village or a tribe) may change quickly.

Genetic and epigenetic transfers are also responsible for creating information interfaces between genes and memes. Issues of evolutionary psychology, like cognitive, affective, and moral abilities, might also be seen as consequences of the coevolutionary dynamics of genes and memes, where culture influences selection for genetic patterns and diverse pools of human genomes enable successful biological and cultural adaptation [20]. The evolution of speech physiology and face communication is a striking illustration of how genes and culture can coevolve in a natural and infospherical way. Genetic modifications that make speech easier to produce have been rewarded by the growing social significance of communication in human society. Early humans developed new areas of the motor cortex to help with speech production. The development of speech reflects a coevolution with the technosphere and sociosphere, and forming speech communities as demes, influences the evolution of different socio-biological communities, as well. This way, coevolution shapes the features of the particular M-E-Os.

In the era of genomics, Richerson and Boyd [21] provided a fresh perspective on the coevolution of genes and cultures. Various memetic areas, including the scientific infosphere, the technosphere, the sociosphere, and the environment itself, are all included in this framework. Their paradigm encourages us to update our interpretative framework and provides flexibility for the M-E-O notion to be adapted to coevolutionary discourse. Genes typically change more slowly than culture does, resulting in unique contexts that subject genes to fresh forms of selection pressure. As a result of the altered contexts that cultural innovations have generated, many human genes that have been demonstrated to be under recent or current selection are changing. One change caused by the rise of agricultural subsistence is the spread of sickle cell and G6PD deficiency genes that protect against malaria in local farmers who live in endemic zones of wetland areas full of insects that spread malaria. It is also conceivable that genetic change is happening in reaction to the unique social context of modern societies. In a similar way, some mother-child rearing practices might cause early traumatic experiences that influence adult attachment styles to significant others, including persons, organizations, or cultural patterns, leading to diverse social epigenetic consequences such as mothering style and stress-biological resilience [22].

4. Coevolution in the context of M-E-O and ANT models

4.1 Contributions of neoevolutionary anthropology

At the heart of the model lies niche construction, whereby organisms not only adapt and evolve to their environment but also actively shape and modify it to suit their needs. This modification of the environment feeds back into evolutionary processes, leading to coevolution between the organisms and their natural, social, technological, and infospherical environments. The M-E-O model, which stands for the combination of material, energy, and information flows that shape the interactions between living systems and their environment, represents the coevolutionary processes between organisms and their environment. Actor-Network Theory also provides a lens through which to view the material, energy, and information flows and understand how they interact to shape the co-evolutionary processes between organisms and their environment. Latour waves a mutual net between the actor and the natural, social, and infospherical environment giving meaning to agency, in a way he describes: "*Three resources have been developed over the ages to deal with agency. The first one is to attribute to them naturality, and to link them with nature. The second one is to grant them sociality, and to tie them to the social fabric. The third one is to consider them as a semiotic construction, and to relate agency with building of meaning.*" [9].

Extending the ANT model, in the M-E-O system anything can be an actor, from individual organisms to physical structures, cultural practices, or social forms. If we consider different M-E-O network patterns as actors and subjects of competitive evolutionary dynamics, it is worth applying historical insights offered by neoevolutionary schools of social and cultural anthropology.

Archeologist V. Gordon Child defined the great turning points of cultures, the Neolithic, Urban, and Industrial Revolutions, and lent primacy to technological inventions, arguing that the "*so-called spiritual results of man's historical activity are in the long run determined by the material forces.*" His stages of social evolution might support our model as shifting from nature-dominated M-E-O through Sociocracy to Technocracy, where technological memetic inventions/mutations are shaping the memetic world of ideas and social forms. In his 1951 work, *Social Evolution,* he recognized "correlations" between sociopolitical institutions and techno-economic stages, but emphasized the significance of cultural divergence and convergence, showing how diffusion and assimilation of innovations take place as societies adapt cultural complexes to the requirements of the local environments. It was the neo-evolutionist turn in cultural anthropology by Leslie White and Julian Steward that shed more light on technological adaptation to environmental challenges and the natural environmental pressures humans had to face. Their new foci, natural determinism and technological possibilism, generated a modified view of culture as adaptive environmental gameplay. They uncovered how the localized social and ecological adjustments of cultures to their own unique environments lead to cultural diversity. Leslie White, the neo-evolutionist anthropologist, clearly articulated the multidimensional character of the culture, differentiating the technological, social, and ideological spheres from each other. White emphasized the determining role of technology, which depends on the effectiveness and volume of "harnessing" and transformation of energy, which he formulated as P (cultural product) = E (energy consumption) x T (Energy utilization). This formula may explain the inequality and selection disadvantage of competing M-E-Os, like in the case of autark solar-based, draught farming agriculture and its industrialized, and chemized fossil-energy-based counterpart. (On the other hand, the issue of sustainability, environmental risks, and benefits might get us to think that, in the long term, at different demographical scales, a solar energy-based agricultural M-E-O is more viable.)

White's three components of culture: technological, sociological, and ideological in a human ecological, and environmental sense, create the three spherical poles of M-E-O. This general theory was particularized by the cultural ecological considerations of Julian Steward. Shifting from the technosphere to the natural environment, we slightly move from possibilism to determinism. The natural environment, including climatic, biomic, and geographical factors, exerts a deep impact on the subsistence system of a given culture. Julian Stewart's cultural ecology offers a proper framework to understand the adaptive efforts of enabling sustainance and identic replication of the given culture in particular environmental conditions. In the context of our M-E-O model, these diverse and particular patterns of adaptation include technospherial, infosperial, memetic, and sociostructural elements in the given determining natural environment. The diverse environmental challenges let Steward draw a picture of "multilinear" cultural evolution based on the various ways of adapting to different environments. Steward considered technology and economics as decisive factors determining the development of a given culture, and political systems, ideologies, and religions were understood as secondary products in an evolutionary frame.

These different neoevolutionary schools can be brought together in a four-dimensional M-E-O framework of interacting natural, technospherical, infospherical, and sociospherial niches, or hybridized coevolutionary agents. In this way, competing M-E-O patterns of different technocultural human-ecological systems, like slash-and-burn horticulturists, agricultural cultivation systems based on hydro elevation technology, and autopoietic systems of solar energy-based cultivation versus allopoietic fossil energy-based industrialized agriculture, show how important the dynamics of energy transformation are in the struggle for survival of different M-E-O formations.

4.2 M-E-Os are subjects of evolutionary competition in the framework of General Theory of Evolution

The human ecological sphere of competing M-E-Os may be seen as the theater of coevolutionary drama. This might be understood as a historical process of economic and social evolution. Following the logic of the general evolutionary theory of Csányi and Kampis, the evolution of M-E-Os as more or less synchronized replicative systems needs a framework, where natural, technological, social, and infospherical elements are the sum of interacting components, that are assembled and disassembled permanently by component-producing processes. There is an energy flow through the M-E-O system capable of exciting some of its building elements. All M-E-Os may be seen as component systems, using the terminology of Csányi and Kampis, where the number and type of building parts of the M-E-O and the energy flux constitute the parameters of the system. Adaptability and sustainability of a given M-E-O rest on the ability of the techno-, info-, and socio-structural compartments to enhance their common replication and survival in changing environmental challenges. The zero system level of M-E-O is an early phase where there are no developed functions compared to later developed, structured complex systems (local, isolated tribes or farms versus nation-states, imperiums, or federations; hallucinations, baloney, and primitive belief systems versus canonized, dogmatic systems of ideas).

We can distinguish two kinds of arrangements of building blocks, which describe the status of components. The parametric information describes the structural information of the given M-E-O manifesting merely the systems' parameters, while the functional information expresses the information related to the arrangement and functions of the building blocks of M-E-O in cooperation, support, defense, memory, anabolic and katabolic processes, cultivation and harnessing of resources, etc.

The components' functional information is also structural information for the M-E-O system. M-E-O systems can last for a long time if they can reproduce, which happens when information is passed down from one generation to the next through genetic and memetic means. Replicative function, as special structural information in M-E-O, denotes the probability of the identical genesis of the same set of components. The rules of relationships serving genetic replication in the socio-sphere are incest taboo, endogamy, and exogamy. The archivation processes in the infosphere by symbolic agency—oral traditions, rituals, writing, canonization, protocols, and technological know-how—also reflect the importance of replicative function as structural information of the M-E-O. We can distinguish between temporal and spatial replication. Temporal replication provides continuous renewal of a system with subsystems over time, conserving the unity and identity of the system. Ritual communication in organizations is a good example of their sustainability over time. Spatial replication generates the replica, which becomes separated in space. Cultural diffusion, fissioning, colonialization, and globalization are examples of the latter. The M-E-O model might be replicated through mass migration, missionary agency, soft power, global media, or other cultural transfers. This replication process may be identical or non-identical, depending on the degree of fidelity.

The autogenesis of the M-E-O system, similar to molecular, or biological genetic replicative systems, is emerging from autogenetic system-precursors, which might be genetically and/or memetically reproduced sets of natural, technospherical, memetic, and social components. According to the evolutionary theory of Csányi and Kampis, it must contain *"at least one cyclic component-producing process, and at least one of the components participates in the cycle, which can be excited by the energy flux flowing through the system"* [13]. The given M-E-O implies human biological, social, and technocultural precursors. For instance, a hunter-gatherer group has simple social rituals, low-level technospherical "extensions" of human capabilities, and a low level of energy transformation compared with agricultural cultivators, industrial, or post-industrial societies. Herbert Spencer, the pioneer of the unilinear evolutionist view of societies, focuses on the growth or progress of complexity. During the process of autogenesis, the M-E-O's functional information builds up, and its replicative information grows as more supercycles branch out and connect to the M-E-O's core formations. The growing circles of differentiation and cooperation (e.g., division of labor) lead to the formation of communities of simultaneously replicating compartments. When the identical replication stage of M-E-O is stabilized, its replication continues as long as the natural, social, technological, or ideological environment does not change. In such a catastrophe-free climax state, M-E-O becomes an autonomous self-maintaining unit, a network of component-producing processes (in the sense of Actor-Network Theory also), which, through the functional interaction of components, produces the same M-E-O pattern that originally produced it. As Csányi and Kampis write, this autogenesis is the evolution of active self-construction, and let us paraphrase, active self-construction of the given M-E-O.

M-E-O has four evolving frontiers, all of which are open to environmental shifts and catastrophes. These natural, technospherical, infospherical, and social metacomponents themselves have different immanent replicative networks that show genetic, memetic coevolutive relationships. Mating strategies, choices, preferences, and reproductive habitus determine genetic reproductive dynamism in a given population under the influence of memetic patterns like socioeconomic conditions, career strategies, cultural capital and habitus, gender ideology, culture, etc. On the other hand, there is a demographic competition between different sociocultural M-E-Os with different religious commitments and identifications, monogamic or polygamist family structures, and socioeconomic and educational standards, even in the same human geographic locality.

The M-E-O system's memetic parts are made up of learned, functionally related patterns of behavior and the physical, technological, and informational things that arise from them. The natural basis of these processes is rooted in the brain, as the behavioral patterns are generated by a population of acquired neural codes in the brain, as a result of gene-culture coevolution [21].

The memetic parts of M-E-O share the general characteristics of evolving, replicative structures in which ideas are representations of learned codes that lead to thoughts, actions, and artifacts that can be shared and copied. This population of memes or ideas contains structural information that might influence the probability of the genesis of other ideas. This memetic system might transform into a (canonized) identical replicative pattern, a system of dogmas, or, under the influence of other memetic impulses, innovations such as memetic mutations, heresies, or a scientific revolution might destroy the claims for the former identical replication of former memeplexes. Natural catastrophes might change cosmology, basic assumptions, ideologies, and values. This happened after the Great Plague changed the European cultural, religious, and ethical "memetic climate" in the XIVth Century. Similar changes are transforming our postindustrial consumerist society, facing the threats of climate change, the "hot world" of the greenhouse climate crisis, or the COVID-19 pandemic. The offer of the so-called Great Reset is a kind of elitarian sociomemetic response based on high-tech infospherical innovations and efforts to change the industrial-post-industrial M-E-O. As memetic parts of an evolving, interactive network of M-E-Os, the ideas could affect the sociosphere, the technosphere, and the infosphere, which could have a big effect on the fourth part, the natural spheres, both outside and inside. Transhumanism is such a radical, transformative ideology to humanize machines and technologize the human being.

Info-technosphere interfaces generate a transfer from brain codes, and memetic patterns to make artifacts, machines, buildings, art, and other material objects. Memetic interfaces transfer replicative structural information into the sociosphere, as social systems are also influenced by systems of ideas. Neural networks direct social perceptions and behaviors, which are molded by imitation and learning. The internal and external populations of ideas as memetic codes regulate social transactions, structuring roles and social groups, too. Social institutions are maintained through ritual communication based on basic assumptions and values, group memories of institutional history, and narratives—one word: populations of ideas.

The infosphere also shares the features of replicative networks in general. Memetic patterns of brain codes might become a coherent replicating population of ideas through language, writing, and mass communication. The networks of ideas that copy themselves are made up of idea compartments that have both structural and functional information. This creates a component system that is similar to chemical, biological, and social component systems. The complexity of idea compartments develops from the status of diffuse, and diverse parametrical information patterns toward a regulated network of functionally connected ideas, where functional, replicative information content is transmitted through memory and transgenerational transfer and in space through all forms of communication: education, propaganda, PR, and mass media, too. M-E-O, with its four-dimensional replication compartment system, has two important features: stability and resilience. Stability is rooted in the ability to keep M-E-O's characteristic features within a given interval of environmental limits through cybernetic processes, while resilience reflects its ability to recover after exposure to the pressure of change.

In the next parts of the paper, we deal with the replication of M-E-O, through the coevolutionary aspects of language based on genetic/memetic abilities and conditions; examining the circular logic of birth and childrearing customs embedded in their natural, technospherical, infospherical, and social milieus. The third focus points to the resilience of M-E-O in the case of the extreme "catastrophic" challenge of the COVID-19 pandemic.

4.3 Language is the house of being

Heidegger [23] wrote the above statement about human existence in his "*Letter on Humanism,*" 1947, which might gain coevolutionary meaning in the context of the M-E-O model. Geertz's statement has similar coevolutionary significance, saying Man "*is an animal suspended in webs of significance he himself has spun*" [24]. According to the hypothesis of Edward Sapir and Benjamin Whorf [25, 26], the semantic structure of a language determines how the speaker's conceptions of the world are formed, and a similar basic concept is the core of Sándor Karácsony's [27] theory, who created a social psycholinguistic theory in the early forties comparing Hungarian and German grammar.

Bourdieu [28] analyzes the power of memes to shape social relationships and dominance. Ideologies and political discourses also shape social relationships, hierarchies, and the ingroup and outgroup positions of social players. Baudrillard writes about the political economy of signs, saying: "*Mastery of the transubstantiation of economic exchange value into sign exchange value is based on a monopoly of the code*" [29]. One can follow how discourses as social memetic products generate institutions and social power dynamics by reading Foucault's discourse theory. These theories have a common basis: memetic determinism, but language also represents and expresses changes in the technosphere, social environment, and natural environment.

Technological advances generate new terminologies, and the development of communication technologies influences social linguistic phenomena. The technospheric changes played an important role in the evolutionary "synchronization" of diverse local dialects, vocabularies, commonsense, and collective mentalities, as the metaphors of the Gutenberg Galaxy and the Global Village of Marshall McLuhan offer a good sample of technomemetic-social coevolutionary dynamism. The electronic mass media of the "global village" and WEB2 social media lent further power to the English language as a new Latin, or lingua franca, mediating strong lifestyle-shaping influence. According to Manuel Castells [30], the essence of globalization rests in the (techno memetic) networks that structure society: "*The culture of real virtuality, created around an increasingly interactive and audiovisual universe, permeates intellectual images and communication everywhere, integrating the diversity of cultures into a kind of electronic hypertext.*" As these codes influence important personal decisions regarding lifestyle, career expectations, gender identification, marriage, reproduction, and migration, their coevolutionary agency is remarkable.

By looking at the archeogenetics of certain populations and the etymology of their words and phrases, it is possible to track the genetic-memetic coevolutionary changes that have taken place in certain communities over time [31]. The scientific discourse regarding genetic and linguistic coevolution is paradigmatically articulated by the works of Cavalli-Sforza [32], who showed, how languages might create barriers to gene flow between different human populations. (The opposite socio-memetic direction is also true when language mirrors the collision of different groups, like in cases of invasion, subversion, or assimilation.)

Cavalli-Sforza and Feldman [33] established a coevolutionary discourse in linguistic anthropology as part of social and cultural anthropology, referring to *gene-culture coevolution*, dealing with dual inheritance processes of *cultural transmission in an evolutionary context*.

4.4 Coevolution and reproductive dynamism in different M-E-Os

The culture of reproduction is the most intimate interface, where genetic and memetic reproduction are intertwined. Mating and reproduction create a "hot" interface of memetic and genetic fusion of different genetic and memetic pasts of wife and husband, with further physiological, transgenerational memetic consequences for the offspring. This genetic and memetic fusion involves the wider family and, in a wider sense, the background of different M-E-Os, mixing cultural, technospherical, and social memetic contents in the sense of Bourdieu's concepts of habitus, social capital, and field [34]. Different M-E-Os produce various memetic, technical Actor-Network settings for reproduction, including nature-dominated, agricultural sociocratic, industrial technocratic, and postindustrial infocratic M-E-Os. The consumerist, neoliberal, and postindustrial infocratic

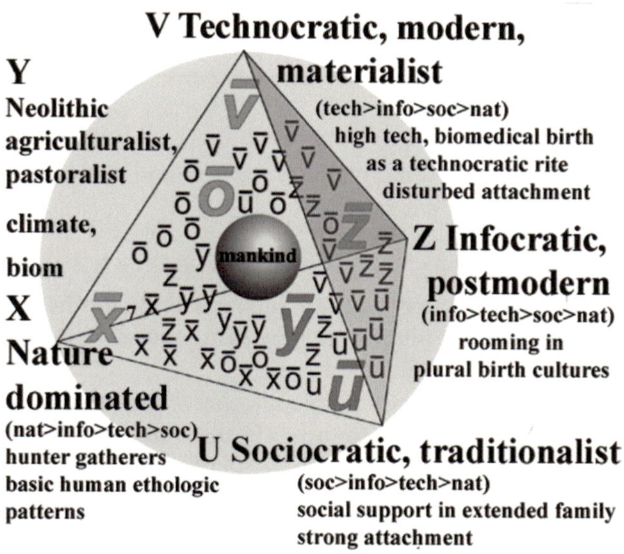

Figure 2.
Historical paradigms of social patterns and childrearing complexes in the MEO context [6].

M-E-Os have a very low reproduction rate (1,2-1,7), not enough to sustain the given population, which leads to economic difficulties and pension crises. Even nations with former family-friendly cultural traditions, like Italy, Ireland, or Central European nations, can experience this. The coevolutionary, genetic, and memetic consequences of reproductive crises of M-E-O in depopulating societies are obvious if mass migration from Asia and Africa is expected to substitute for the loss of given populations. In contrast, residents of the "less developed" agricultural, sociocratic M-E-O desire more children and live in extended families even in poor socioeconomic circumstances.

Examining how infospherical, technological, and social memes affect reproduction is crucial. The condition of human ethological norms in early mother-child attachment systems may be an appropriate target to draw attention to the differences between agrarian, industrial, and postindustrial M-E-Os. The circumstances, laws, and medical procedures surrounding childbirth are influenced by the cultural, technological, and cognitive environment as well as the social and institutional structures of the specific society. It is important to examine the influences of technology, information, and social memes on reproduction, including contraceptives, changing gender roles, counter- or profamily state policies, and various "r" and "K" reproductive strategies [35] found predominant in different social strata. In addition to having an impact on the mother-child bond, birth has long-term psycho-developmental and projective secondary socio-cultural effects (**Figure 2**).

1. **Nature dominated** archaic magical complex: Various kinship systems, archaic group relationships, Traditional Birth and Childrearing Complex, socially embedded material life (Gift systems, reciprocity).

2. **Sociocratic, traditionalist:** Extended family, traditional communities, socially embedded material life, (Gemeinschaft), Traditional Birth and Childrearing Complex, secure attachment ideal.

3. **Technocratic modern materialist:** Nuclear family, institutionalized, rationalized social life (Gesellschaft) Technocratic birth, a detached, disembedded market economy, childrearing complexes, and tendencies toward distorted attachment.

4. **Infocratic, postmodern:** Plural birth cultures, home birth, rooming in, feminist criticism of paternal obstetrics.

The M-E-O context consists of a variety of technological settings, customs, and belief systems. These elements may have an impact on how mothers behave and raise their children, resulting in different internal working models (IWM) of secondary adult attachment styles [36].

Environmental and climatic factors affect mother-child relationships in every culture. For instance, traditional communities in the southern countries usually permitted piggybacking, whereas infants were frequently kept in cradles in the Northern countries. According to Freedman and Gorman, keeping one's distance and cueing for independence between a mother and child were common signs of insecurity in Iceland [37]. Different M-E-Os that are based on pastoralist, agricultural cultivator, or industrial technocultural subsistence systems produce different birth practices, protocols, and attachment organizations.

The intergenerational transcription of attachment styles takes place within the context of the family, and it may even be essential for the repetition of sociocultural M-E-O patterns. Anthropological works by Kardiner, Barry, Margaret Mead, Whiting, and Child illustrate the importance of childrearing in cultural transmission. According to Kardiner [38], secondary institutions like art, religion, and rituals are partially the results of this basic personality structure, while the primary social influences of childrearing practices and specific child-training practices shape the so-called basic personality structure. The comparison of 110 ancient, prehistoric cultures by Barry, Bacon, and Child [39] revealed two fundamental trends: secure mothering and appropriate gender socialization. In 61% of the civilizations, girls showed higher degrees of responsibility, and in 82% of the cultures, they could show superior caring and caregiving skills. Boys, on the other hand, displayed higher degrees of dominance, self-centered dignity, and achievement-centered views in 87% of these cultures under examination. According to Barry, children in shepherding, pastoralist civilizations are nurtured with an emphasis on developing strong independence and competence, in contrast to children in farming cultures, who are taught submission [40]. According to Whiting and Child [41], the same early experiences might give rise to the same unconscious tensions and fixations as neurotic tendencies that are reflected in the cultural belief system. When Margaret Mead [42] observed Mundugumors and Arapesh people, it became clear that child-rearing practices might affect the communication and behavioral norms of interpersonal relationships, ethics, etc. in the given tribes. Comparing East-West cultural differences, separateness against connectivity, individualism versus collectivism, and perhaps detached versus attached motives, more family context distinctions become apparent [43]. The sociocultural and birth culture perspectives taken by LeVine [44] are a good illustration of the four-dimensional human ecological dynamics of M-E-O. His advanced methodology widens the scope of investigations into the historical and intercultural variances in birth and childrearing practices that are rooted in human ecological variables that include geographic, technological, and cultural contexts. Sustainability, the psychological basis of economic attitudes, and environmental attitudes are influenced by attachment styles shaped by mother-child relationships and environmental education; in this way, our environmental crisis is a coevolutionary challenge [45].

Industrial and postindustrial M-E-Os differ significantly from preindustrial M-E-Os in terms of giving birth and raising children. As a result of modernization and industrialization based on new technologies, the time for uninterrupted mother-child attachment systems was reduced to 6 weeks. Due to this depriving human ethological condition, avoidant-independent and anxious-ambivalent attachment tendencies proliferated. Millions of children's attachment systems were burdened or twisted by Hungary's rigorous, technocratic birth rules and early weaning (after 6 weeks) in the 1950s and 1960s. When this generation experiencing this early burdened attachment reached their age of thirties and fourties, sociologists reported a rising divorce rate, and the reproduction rate decreased. Even in the millennium era, when the children of the baby boomers became fertile, the low reproduction rate persisted. According to Csóka and Lázár [46], who reported a considerable shift toward avoidant, dismissive adult attachment (58% in the representative sample of 600 under age 35) in both genders, it was a transgenerational impact, where significant relationships between avoidants and the aversion toward having children were confirmed, too.

A critical interpretive discourse of medical anthropology [47] vividly illustrates the issues with technological birth with coevolutionary accents. Based on human ethological considerations, the traumatic separation of the newborn from the mother in the first hours after birth during the imprinting phase, or keeping the baby apart from the mother between breastfeeding sessions can be seen as stressful micro-traumas. For instance, rats can suffer separation traumas in just 180 minutes, with long-lasting neurophysiological repercussions and epigenetic consequences [48].

As a remedy, "rooming in" is commonly used in maternal wards, occasionally replicating a homely birth setting with the father's involvement. Changes in employment and professional expectations and limitations that force some women to leave their newborns in the care of daycare providers during the first few months of life further strain the mother-child bond.

The changing value and role system caused by industrialism, and the postmodern information society that followed, as well as changes in female life strategies and gender identity, childbearing attitudes, all contributed to the present day's coevolutionary M-E-O shift with considerable demographical and cultural impact.

4.5 COVID-19 and M-E-O

Whether a disaster physically damages networks or has an infospherical impact, which results from active information carriers interfering with the social system's cycles of reproduction, it can cause functional changes that disrupt or destroy some M-E-O. Some compartments may lose their functional relationships and stop participating in the replicative cycles, changing the identity of the system. The COVID-19 pandemic inspired intellectuals to rethink the cosmological, economic, and ethical memeplexes of our modern M-E-Os with suggestions to change their techno- and socio-spheric compartments as the concepts of the Great Reset [49] emerged in mass communications, much like how the Great Plague had profound cosmological, ethical, and social effects on Medieval Europe. The pandemic itself had a significant impact on the techno-economic environment as well as the infosphere portion of M-E-O. The sociosphere also faced significant problems at both the micro and macro levels. The entire M-E-O entered a new "attractor basin," a posttraumatic state of the system, as a result of these circular and reverberating changes, while the global M-E-O showed significant resilience on the one hand with infospherical synchronization and technospherial adaptation, and on the other hand, a tendency toward industrial deglobalization emerged because of breaking off international travel and supply chains.

The global M-E-O system underwent profound, multitargeted modifications as a result of the COVID-19 pandemic. It altered the commonplace social and cultural structures that govern our lives. In an era of the sensual culture of services and entertainment, lockdown, social isolation, internet-based communication, home office work, suspension of travel, theater life, communal cultural life, restaurant business, and hotel business all proved to be prone to by radical socio-cultural transformation driven by epidemiological considerations and policy. The techno-memetic, info-memetic, and socio-memetic compartments all underwent these alterations. The SARS-CoV-19 pandemic has had an impact on the relationships and organizational frameworks of adaptive and maladaptive human agency in the context of ecological processes that create, mold, and sustain people in their multidimensional, intersecting environments. De Ver Dye and Muir's [50] critical medical anthropological approach provides the appropriate perspective for this investigation. This method is quite similar to the M-E-O tetrahedron model's hybridizing notion [6]. Destruction of long supply chains, market competition over the availability of masks, respirators, and drugs like remdesvir and favipiravir, as well as vaccines with various East-West cultural-political connotations (AstraZeneca, Pfizer-BioNTech versus Sputnik V, Sinovac), created info-dynamic tension at the level of health politics and media, as well as among providers of health services and people from various sociocultural backgrounds. The new normal of lockdown, giving up personal freedom for perceived security, and media exposure to the COVID-19 epidemic's deadly hazards all had a significant impact on the psychological, social, and memetic climate of pandemic M-E-O (**Figure 3**).

The COVID-19 outbreak seems to have caused a social shift toward a new, more "cyberized" way of life. Robots will soon replace hundreds of millions of people's physical labor thanks to the Internet of Things and Artificial Intelligence. The challenges of permanent adaptation strain in the world of work and the growing risk of sedentism caused by the virtualization of human work through information technology could make people less able to deal with stress, which could increase biological health risks.

The COVID-19 pandemic had a significant impact on regional and local businesses, such as restaurants, hotels, and family-run operations, and it also created new opportunities for the digitalization of education, homework, and several other jobs where intellectual or other products could be transferred over the Internet. Although the COVID-19 pandemic has affected everyone, the poor and vulnerable groups, such as low-skilled laborers and immigrants, have suffered significantly more than the wealthy in terms of fatalities, infection risk, decreased earnings, and unemployment. Telemedicine, which puts patients and doctors at a distance, was the direction in which health services were moving, but eHealth showed a real surge. The medical sector, Big Pharma, vaccination, and the market for items that boost the immune system were all growing. B vitamins, vitamin C, vitamin D, zinc, magnesium, and multivitamins all rank among the immune enhancers recommended by the WHO, and sales of these supplements have increased significantly. Additionally, certain independent business owners and service providers, such as trainers, coaches, and advisers, expanded their services online. Local manufacturers established their websites, online stores, or other online platforms to enable the ordering of their products

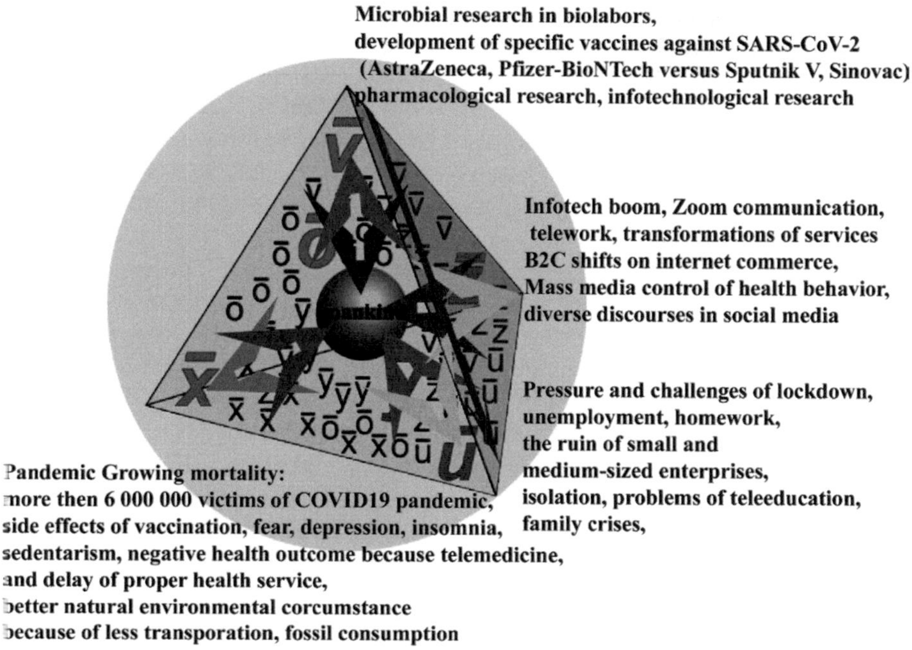

Figure 3.
Effects of the SARS-CoV-19 pandemic on the M-E-O.

through interfaces. Aside from that, the opening of virtually huge department stores and mass producers gave farmers' markets and consumers a chance to help local small companies survive. The most important Hungarian business players also emphasized that the outbreak contributed most to an increase in B2C sales. Not only have the already established courier businesses grown recently but several unimportant new players have also entered the market. Rural areas offered great opportunities for this, such as creating a network in areas where larger businesses only had a limited scope of operations and concentrated on a single product category, such as the goods of a major producer of vegetables and fruits.

The resilience of M-E-Os at different scales during the COVID-19 pandemic was expressed through an articulated memetic, technological, and social transformation, as an adaptive response to a catastrophic challenge, where over 6 million people died and people paid a very high price expressed in health deficits, family situations, employment, and loss of feeling of security.

5. Summary

The coevolutionary model of Man-Environment-Organism focuses on the unity of human-environment and offers a framework for human coevolution, where the given, particular M-E-O is the subject of evolutionary selection. On the one hand, the Darwinian fight for survival is the fate of all forms of M-E-Os, and on the other, Kropotkinian cooperation and mutuality among different M-E-Os can offer an adaptive advantage. We have shown, how human ecological models, some general theories of evolution and coevolution, and neoevolutionary concepts of archeology and cultural anthropology meet the M-E-O model as a common framework.

The human ecological sphere of competing M-E-Os is the theater of the coevolutionary drama, a historical process of memetic, economic, and social evolution. M-E-Os, be they families, nations, or organizations, are synchronized replicative systems with natural, technological, social, and infospherical components. The function of M-E-O depends on the ability of components to influence the probability of the genesis or survival of other components. The sustainability of M-E-O systems depends on their replicative functions, which result from transgenerational information transcription in genetic and memetic ways. Autogenesis of M-E-O systems comes from autogenetic system precursors, which can be sets of natural, technological, memetic, and social parts that are passed down through genes or memes.

We overviewed, how language has a memetic-social, memetic-technological transfer function in Actor-Network agency and coevolutionary processes, too. Language plays an important, niche-constructive role in influencing the selection of genes and memes. Language, ideologies, and political discourses shape social relationships, hierarchies, and the ingroup and outgroup positions of social players. The archeogenetics of given populations and the etymological analysis of their vocabulary and idioms let us trace genetic-memetic coevolutionary changes in the past of given communities.

We examined the culture of biological replication, the birth culture, and mother-child relationships as genetic/memetic transgenerational mixed ways of selective dynamism in M-E-Os. Different M-E-Os create different memetic and technological Actor-Network contexts for reproduction, with different cultural and cognitive worlds and social institutional arrangements influencing the conditions, legal regulations, and technical practices of birth and early mother-child relationships. The mother-child attachment organization is influenced by various technical environments, rituals, and belief systems, influencing maternal behavior and childrearing. Family structures play a crucial role in the intergenerational transcription of attachment styles and the replication of sociocultural M-E-O patterns. Child-rearing practices shape cultural transmission, with some cultures focusing on nurturing and caregiving abilities while others emphasize independence and competence.

We also examined the impact of the environmental catastrophe of the COVID-19 pandemic on the late- or postindustrial M-E-O, altering technological, social, and cultural structures. The pandemic led to a new traumatic "attractor basin" of global M-E-O showing resilience and a tendency toward radical transformation of technospherial and infosperial compartments of the M-E-O, including synchronized infospherial globalization and industrial and commercial deglobalization with broken supply chains. The COVID-19 pandemic has significantly impacted regional and local businesses, causing a decline in gross domestic product and affecting vulnerable groups like low-skilled laborers and immigrants. Telemedicine, eHealth, and the market for immune-boosting supplements have experienced significant growth, and telework and home office activities have expanded. The COVID-19 pandemic has highlighted the importance of understanding sociocultural and psychological networks in public health problems. The Actor-Network dynamism induced by the COVID-19 pandemic proved the resilience, the interrelatedness, and the potential for effective synchronization of the info-, techno-, social-, and natural compartments of M-E-O with a high level of functional and replicative information as an evolutionary system. The coevolutionary frame of linguistic-memetic and population genetic connections, the coevolutionary interrelatedness of memetic and biological components of the human reproduction process, and the interrelated and synchronized changes of the M-E-O during the pandemic as a natural disaster support accepting the M-E-O as the subject of coevolutionary dynamism.

Author details

Imre Lázár
Institute of Social and Communication Sciences, Faculty of Humanities and Social Sciences, Károli Gáspár University of the Reformed Church in Hungary, Budapest, Hungary

*Address all correspondence to: lazar.imre@kre.hu

References

[1] Lázár I. E-K-O-lógia (M-E-O-logics). Országépítő. 1997;**3-4**:26-34

[2] Boulding K. Ecodynamics: A New Theory of Societal Evolution. Beverly Hills, London: Sage; 1978

[3] Csányi V. General Theory of Evolution. Budapest: Hungarian Academy of Sciences; 1982. pp. 128

[4] Latour B. We have never been modern. Cambridge, Massachusetts: Harvard University Press; 1993

[5] Odling-Smee FJ, Laland KN, Feldman MW. Niche Construction: The Neglected Process in Evolution. Princeton, NJ: Princeton University Press; 2003

[6] Lázár I. "Attached Files" Anthropological Essays on Body, Psyche, Attachment and Spirituality. Newcastle upon Tyne: Cambridge Scholars Publishing; 2015

[7] Jablonka E, Lamb MJ. Evolution in Four Dimensions: Genetic, Epigenetic, Behavioral, and Symbolic Variation in the History of Life. Cambridge, MA: MIT Press; 2005

[8] Callon M. The sociology of an actor-network: The case of the electric vehicle. In: Callon M, Law J, Rip A, editors. Mapping the Dynamics of Science and Technology. London: Macmillan Press; 1986. pp. 19-34

[9] Latour B. On actor-network theory: A few clarifications. Soziale Welt. 1996;**47**(4):369-381

[10] Law J. Notes on the theory of the actor-network: Ordering, strategy, and heterogeneity. Systems Practice. 1992;**5**:379-393

[11] Kropotkin P. Mutual Aid: A Factor of Evolution. (1905). Boston: Extending Horizons Books, Porter Sargent Publishers; 2005

[12] Guimarães PR Jr, Jordano P, Thompson JN. Evolution and coevolution in mutualistic networks. Ecology Letters. 2011;**14**:877-885

[13] Csânyi V, Kampis G. Modelling biological and social change dynamical replicative network theory In: Ervin L, editor. The New Evolutionary Theory. New York: Gordon and Breach Science Publisher; 1991. pp. 77-93

[14] Domingo E. Interaction of Virus Populations with Their Hosts. Virus as Populations. 2016:123-168

[15] Thompson JN, Cunningham BM. Geographic structure and dynamics of coevolutionary selection. Nature. 2002;**417**:735-738

[16] Dawkins R. In: Hofstadter D, Dennett D, editors. The Selfish Gene (1976) in the Mind's I. New York: Basic Books; 1981

[17] Blackmore SJ. The Meme Machine. Oxford: Oxfords University Press; 1999

[18] Velikovsky JT. The Holon/patron theory of the unit of culture (or the meme and Narreme) in science, media entertainment and the arts. In: Connor A, Marks S, editors. Chapter in Creative Technologies for Multidisciplinary Applications. New York: IGI Global; 2016

[19] DDiCarlo CW. How Problem Solving and Neurotransmission in the Upper Paleolithic Led to the Emergence and Maintenance of Memetic Equilibrium in

Contemporary World Religions Politics and Culture; 27 April 2010;**2010**(1)

[20] Gintis H. Gene–culture coevolution and the nature of human sociality. Philosophical Transactions of the Royal Society of London. Series B, Biological Sciences. 2011;**366**(1566):878-888

[21] Richerson PJ, Boyd R. Not by Genes Alone How Culture Transformed Human Evolution. Chicago: University of Chicago Press; 2004. p. 342

[22] Lázár I. At the cradle of psychobiological risks: Distorted attachment organizations in human ecological and evolutionary contexts. In: Watson P, editor. Social Behavior Evolutionary Pathways, Environmental Influences and Impairments. New York: Nova Publisher; 2014

[23] Heidegger M. Über den Humanismus. Vittorio Klostermann, Frankfurt am Main; 1949

[24] Geertz C. The Interpretation of Cultures. New York: Basic Books; 1973. p. 5

[25] Sapir E. Language: An Introduction to the Study of Speech. New York: Harcourt, Brace and Co.,; 1921

[26] Whorf BC. Language, Thought, and Reality: Selected Writings of Benjamin Lee Whorf. In: John B. Carroll, editor, MIT Press. New York, London: John Wiley & Sons, Inc.; 1959

[27] Karácsony S. Magyar Nyelvtan társas-lélektani Alapon (Hungarian Grammar in Social-Psychological Context). Budapest: Bethánia; 1938

[28] Bourdieu P. Language and Symbolic Power. Oxford: Polity Press; 1991

[29] Baudrillard J. Pour une critique de l'économie politique du signe. Paris: Gallimard; 1972

[30] Castells M. The culture of real virtuality: The integration of electronic communication, the end of the mass audience, and the rise of interactive networks. In: The Rise of the Network Society: With a New Preface. Vol. 1, 2nd ed. Hoboken, New Jersey: Wiley; 2009

[31] Lázár I. Etnoökológia és nyelvtörténet (Ethnoecology and history of languages) In: Ferenczné, Szőcs É, Gazda J, editors. A magyarság rétegei (Layers of Hungarians). Covasna: Körösi Csoma Society; 2020. pp. 229-251

[32] Cavalli-Sforza LL. Genes, Peoples, and Languages. New York: North Point Press; 2000

[33] Cavalli-Sforza LL, Feldman M. Cultural Transmission and Evolution. Princeton: University Press, Princeton; 1981

[34] Velikovsky JT Creative Practice theory. Ev. Cult StoryAlity. [PhD dissertation] 2012

[35] Belsky J, Steinberg L, Draper P. Childhood experience, interpersonal development, and reproductive strategy: An evolutionary theory of socialization. Child Development. 1991;**62**:647-670

[36] Bowlby J. Attachment: Attachment and Loss. Vol. 1. New York: Basic Books Classics; 1969

[37] Freedman GD, Gorman J. Attachment and the transmission of culture an evolutionary perspective. Journal of Social and Evolutionary Systems. 1993;**16**(3):297-329

[38] Kardiner A, Linton R, DuBois C, West J. The Psychological Frontiers of Society. New York: Columbia University Press; 1945

[39] Barry H, Bacon MK, Child EL. A cross-cultural survey of some sex

differences in socialization. Journal of Abnormal and Social Psychology. 1957;**55**:327-332

[40] Barry H et al. Relation of Child training to subsistence economy. American Anthropologist. 1959;**61**:53-63

[41] Whiting JWM, Child IL. Child Training and Personality. A Cross-Cultural Study. New Haven and London: Yale University Press; 1935

[42] Mead M. Sex and Temperament in Three Primitive Societies. 1st ed. New York: HarperCollins Publishers Inc; 1962

[43] Takeshi T, Lau A. Connectedness versus separateness: Applicability of family therapy to Japanese families. Family Process. 1992;**31**:319-340

[44] LeVine RA, Levine S, Dixon S, Richman A, Leiderman H, Keefer CH, et al. Child Care and Culture. Lessons from Africa. Cambridge: Cambridge University Press; 1996

[45] Lázár, I. Mothers and mother nature: Attachment, detachment and human ecological integrity In: Crabbé P. szerk. Implementing Ecological Integrity. Restoring Regional and Global Environmental and Human Health (NATO Science Series: IV. Earth and Environmental Sciences). Dordrecht: Kluwer Academic Publishers, 2000; 249-260

[46] Csóka S. Lázár. Kötődési jellemzők a mai magyar népesség körében; a kötődési biztonság mint potenciális védőfaktor. (Attachment factors in contemporary Hungarian population... In: Susánszky Éva, Szántó Zsuzsa, editors. Magyar Lelkiállapot. 2013; 268 p. Budapest: Semmelweis Kiadó és Multimédia Stúdió; 2013. pp. 187-199

[47] Davis-Floyd R. The technocratic body: American childbirth as cultural expression. Social Science & Medicine. 1994;**38**:1125-1140

[48] Meaney MJ, Diorio J, Francis D, Widdowson J, LaPlante P, Caldii C, et al. Early environmental regulation of forebrain glucocorticoid receptor gene expression: Implications for adrenocortical responses to stress. Developmental Neuroscience. 1996;**18**(1-2):61-72

[49] Schwab K, Malleret T. COVID 19 The Great Reset. Geneva, Switzerland: Forum Publishing, World Economic Forum; 2020

[50] De Ver Dye T, Muir E, Farovitch L, Siddiqi S, Sharma S. Critical medical ecology and SARS-COV-2 in the urban environment: A pragmatic, dynamic approach to explaining and planning for research and practice. Infectious Diseases of Poverty. 2020;**9**(1):71

Section 2

Bioinformatics, Immunogenetics and Vaccines

Chapter 2

Recent Advancement on In-Silico Tools for Whole Transcriptome Analysis

Vidya Niranjan, Lavanya Chandramouli, Pooja SureshKumar and Jitendra Kumar

Abstract

Delving into the intricate world of transcriptome analysis, this chapter unfolds the story of gene expression in organisms. The classic DNA microarray and RNA-seq methods have long been the pillars, with RNA-seq taking the spotlight for its superior resolution in understanding dynamic aspects. Yet, tools like Hisat2 and DESeq2, while effective, come with the drawback of being time-consuming and reliant on powerful GPUs. The need for quicker, less resource-intensive techniques has sparked a shift toward simpler R and Python-based tools that not only sidestep GPU dependence but also offer enhanced graphical representations. As we navigate through the content, the chapter draws a vivid comparison between the established tools and the emerging ones, highlighting the pressing need for innovative approaches in transcriptome analysis. The narrative guides readers through the fundamentals, from the Central Dogma's backstory to the pivotal role of RNA in gene expression and disease. It uncovers the nuances between RNA-Seq and microarray technologies, providing a comprehensive overview of tools for data collection and interpreting changes in gene expression. Our journey extends to the latest breakthroughs, such as the TACITuS platform and the TALON pipeline, tailored for in-depth analysis of transcriptomes using long-read data. The chapter concludes by emphasizing the ever-growing significance of transcriptomics in unraveling complex biological phenomena, with a spotlight on the promising applications of next-generation sequencing. A comprehensive summary ties it all together, detailing the step-by-step protocol of transcriptome analysis, along with insights into current tools, their advantages, and limitations, providing readers with a holistic understanding of their practical application and outcomes.

Keywords: transcriptome analysis, in-silico tools, current trends, run time, memory, protocol

1. Introduction

In 1950, Watson and Crick introduced the Central Dogma, outlining the directional flow of genetic information within cells [1]. This fundamental principle of molecular biology involves two key processes: the transcription of DNA into RNA and the translation of RNA into proteins [2]. Subsequent research revealed the existence of various RNA types, including messenger RNA (mRNA), transfer RNA (tRNA), and ribosomal RNA (rRNA), playing pivotal roles in the synthesis of proteins [3]. During the 1980s, investigations into U-rich RNA, particularly in *Tetrahymena thermophila* and bacterial RNase P complexes, revealed the catalytic function of RNA akin to ribosomes. Subsequent findings highlighted the existence of micro-RNA (miRNA) and their regulatory properties. Both messenger RNA (mRNA) and non-coding RNA (ncRNA) play pivotal roles in governing gene expression and participating in cellular development [4]. RNA stands out as a crucial macromolecule in biological cells, transcribing essential messages from DNA to facilitate protein synthesis, thereby sustaining life. Minor alterations in transcription can disrupt the entire mechanism, sometimes leading to severe diseases. The transcriptome, encompassing all messenger RNA molecules (mRNAs) expressed by an organism, characterizes the intricate web of genetic information. Additionally, the term "transcriptome" extends to describe the mRNA transcripts within a specific cell or tissue type. The field of transcriptomics closely examines the regulation, variation, and mechanisms governing RNA molecules in cellular processes [5]. Comprehensive examination of the entire transcriptome has become increasingly vital for comprehending the modified expression of genetic variants implicated in complex diseases such as cancer, diabetes, and cardiovascular conditions. Primarily, transcriptome analysis unveils fresh insights into biomarker exploration, establishing gene-centric benchmarks for personalized medicine and therapies [6]. Furthermore, transcriptome analysis plays a significant role in advancing research on long-term effects of COVID-19 [7]. By scrutinizing genome-wide differential RNA expression, researchers can gain a deeper understanding of the biological processes and molecular mechanisms governing cell fate, development, and the progression of diseases.

The analysis of the transcriptome commonly involves a comparative evaluation between two groups, notably healthy and diseased conditions [8]. This approach proves valuable in elucidating the functionalities of genes and regulatory pathways. Advanced techniques in transcriptome analysis encompass microarrays, which

	RNA-Seq	**Microarray**
Definition	RNA-Seq (RNA-Sequencing) can examine the quantity and sequences of RNA in the given samples using Next-generation sequencing (NGS)	Microarray detects the expression of genes.
Technique	Sequence-based	Hybridization-based
Sequence	Identify novel RNA sequences	Identify known sequences
SNP	Identify SNP except new SNPs for low abundance	Cannot identity SNPs
Sensitivity	Highly sensitivity	Comparatively low sensitivity
Accuracy	High accuracy	Low accuracy
Labor intensity (Sample preparation and data analysis)	High	Low
Technically reproducibility	>99%	>99%

Table 1.
Comparison of microarray and RNASeq technologies.

provide a comprehensive quantitative and qualitative gene expression profile of the sample by scrutinizing the entire transcriptome. Additionally, RNA-Seq employs high-throughput sequencing to capture all sequences, offering an enhanced perspective on the complexity levels within the eukaryotic transcriptome [9]. **Table 1** illustrates distinctions between RNA-Seq and Microarray technologies.

Despite the drawbacks of RNA-Seq, such as high costs and the need for powerful computing systems, this technology offers the advantage of not depending on prior sequence information. It facilitates the exploration of both known and unknown transcripts. The RNA-Seq process entails preparing libraries and sequencing RNA samples through diverse platforms like Illumina, Nanopore, and PacBio. The sequencing techniques yield fastq files, and the subsequent bioinformatics pipeline is implemented using suitable tools. The following sections delve into discussions about the existing tools, emerging technologies, and the protocol involved in RNA-Seq.

2. Glimpse of protocol and tools used for transcriptome analysis

2.1 Data collection and processing

Data collection is a key process for protocol development, decision making, planning and research. Effective data collection of the transcriptome sequences gives the required outcome and enables to predict future trends. Quality of the reads plays a major role in the analysis. To identify sequencing errors, PCR artifacts, or contamination, Analyzing the sequence quality, GC content, adaptor presence, overrepresented k-mers, and duplicate reads is part of quality control for the raw reads. Acceptable duplication, k-mer, or GC content levels vary organism- and experiment-specific, although these values should be uniform among samples in the same experiments. With above 30% disagreement, we advise eliminating outliers. In contrast to NGSQC [10], which can be used on any platform, FastQC [11] is a popular tool for carrying out these analyses on Illumina reads. In general, read quality declines at the 3′ end of reads, and if it drops too low, bases should be deleted to increase the mapping quality. The low-quality reads can be removed using Trimmomatic and FastX toolkit [12, 13].

2.2 Denovo assembly

One or more contigs are created from partially or completely overlapping reads and one or more scaffolds are created by joining groups of contigs that are overlapping or non-overlapping. A single chromosome is created by joining together groups of overlapping or non-overlapping scaffolds. Reads must overlap by a certain number of base pairs, called k-mers, in the contig assembly process before they can be mapped together. Contigs can be connected to one another during the scaffold assembly process without necessarily overlapping. This can be attributed to paired-end sequencing. Scaffolds are connected via a gap-filling, gap-closing, or genome-finishing procedure during the chromosomal construction stage. Using only short-read technology, it can be challenging and occasionally impossible to finish this last stage. Despite considerable improvement in this area, the presence of repetitive sequences can prevent gap-filling using only short reads [14]. The ability to generate large amounts of RNA-sequencing data has led to the creation of a variety of reference-based and de novo transcriptome assemblers, each of which has unique benefits and drawbacks. The only organisms that can use reference-based methodologies are those with entire, well-annotated genomes, even though many transcriptome investigations routinely use them. De novo transcriptome reconstruction from short reads is challenging, and this challenge is exacerbated by alternative splicing, paralogous genes, and the diversity of gene expression levels. rnaSPADES and Trinity [15, 16] are one of the robust tools for performing De novo assembly.

2.3 Read alignment

Typically, reads are mapped to either a genome or a transcriptome. The percentage of reads that are mapped, which serves as a general indicator of sequencing accuracy and the presence of contaminating DNA, is a crucial measure for determining the mapping quality. In this regard, we forecast 70 and 90 percent of typical RNA-seq readings to map onto the human genome [17], with a sizeable portion of reads mapping to a small number of identical areas equally well (referred to as "multi-mapping reads"). The uniformity of read coverage across exons and the mapped strand represents crucial parameters. If there is a predominant accumulation of reads at the 3′ end of transcripts in poly(A)-selected samples, it could suggest suboptimal RNA quality in the initial material. The standard tool for alignment: HISAT2 The reads from RNA sequencing studies can be aligned using the extremely effective approach known as HISAT (hierarchical indexing for spliced alignment of transcripts). HISAT utilizes dual types of indexes for alignment: a comprehensive whole-genome FM index for initial alignment anchoring, and several localized FM indexes for rapid extension of these alignments. This indexing methodology is grounded in the Burrows-Wheeler transform and leverages the Ferragina-Manzini (FM) index. Each of the 48,000 local FM indexes in HISAT's hierarchical index for the human genome represents a genomic area of about 64,000 base pairs [18].

2.4 Quantification

The primary use of RNA-seq commonly involves assessing transcript and gene expression levels. This application heavily relies on the quantification of reads aligning to each transcript sequence, although alternatives like Sailfish and others may involve counting k-mers in reads without the need for mapping [19]. This gene-level quantification approach, which quantifies genes rather than transcripts, usually disregards multireads and uses a gene transfer format (GTF) file including the genomic coordinates of exons and genes. Raw read counts cannot be used to evaluate expression levels between samples since they are influenced by factors such as transcript length, total reads, and sequencing biases. The feature-length and library-size influences will be eliminated using the within-sample normalization method RPKM (reads per kilobase of exon model per million reads) [20]. This metric and its subsequent modifications, such as FPKM (fragments per kilobase of exon model per million mapped reads) – a measure of normalized transcript expression within a sample like RPKs – and TPM (transcripts per million), are the frequently reported values for RNA-seq gene expression. It's important to highlight that for single-end (SE) values, FPKM and RPKM are interchangeable, and TPM can be derived from FPKM using a straightforward method.

2.5 Differential gene expression

Conducting differential expression analysis necessitates the comparison of gene expression values across different samples. However, RPKM, FPKM, and TPM normalization methods mitigate the impact of a crucial factor for inter-sample

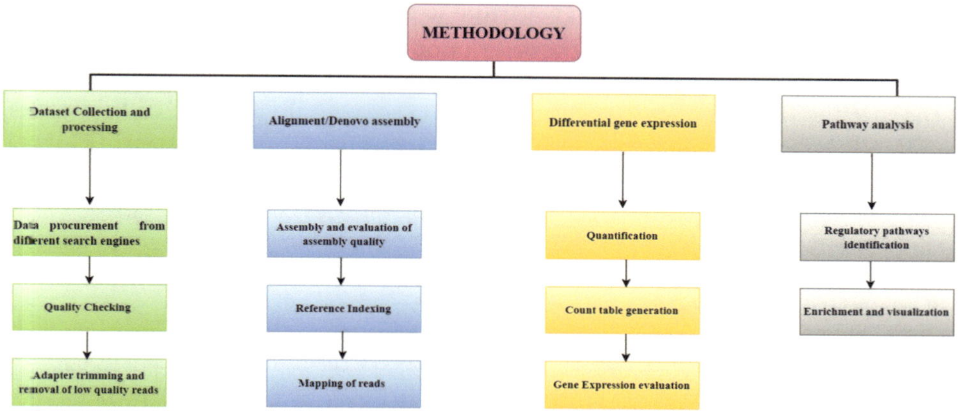

Figure 1.
Protocol for computational transcriptomics: the figure is inclusive of all the steps in transcriptomics; dataset collection, alignment, assembly, differential gene expression and pathway analysis.

comparisons – sequencing depth. Whether directly or by considering the number of transcripts, which may vary substantially between samples, these methods normalize based on total or effective counts. Notably, they exhibit suboptimal performance in cases where samples display heterogeneous transcript distributions, meaning that highly and differentially expressed features can distort the count distribution [21]. One of the important parameters by which the Differential Gene Expression is evaluated is log2fold change value. Negative and positive fold change values are taken into consideration. Negative indicates downregulation of the genes and positive value indicates the upregulation of genes. The most well studied tools for differential gene expression would be DeSeq2, Ballgown, Cuffdiff and Cufflink [22–25]. **Figure 1** shows the complete protocol involved in Computational Transcriptomics.

3. New era for transcriptome analysis

3.1 Dataset collection

3.1.1 Gene expression omnibus

The Gene Expression Omnibus (GEO), managed by the National Centre for Biotechnology Information (NCBI), is a leading repository for high-throughput genomics data. Let us talk about GEO. It's a hub for gene expression studies and is fully accessible to researchers everywhere. Everyone can give their contributions, making the database ever-changing and rich with variegated test data. Searching through GEO is smooth thanks to its user-friendly interface. You can easily download any dataset, detailed metadata, and more. But GEO is more than gene expressions. The repository also keeps other omics data. It follows community standards. This means that everyone can use the data, promoting easy sharing and use of information. GEO is critical in genomics research progress, providing learning tools, backing up research work, and aiding in scientific collaboration and discovery.

The Gene Expression omnibus [26] is an International public repository that provides free access to the high throughput gene expression and functional genomics datasets. It is maintained by National Centre for Biotechnology information (NCBI) and is supported by National Library of Medicine (NLM). The raw files in fastq format obtained after sequencing is deposited along with the descriptions, experimental design, attributes, and information on the protocol for study. The database has provision for Direct retrieval of the specific GEO record and quick access to datasets with the appropriate keywords.

The GEO database offers two separate search engines: (i) GEO datasets, (ii) GEO profiles. GEO datasets: The search engine is used for accessing the datasets of specific study. The submitter's platform, sample, and series entries comprise Database which is supplemented with curated gene Expression Dataset records. Every record has an accession code, title, synopsis, species information, and link to relevant data which will result in thorough and meaningful recovery.

By using the GEO Profiles, people can make the task of finding gene expression profile easier. The database holds carefully compiled set records which forms these profiles. The name of the gene, as well as that of the dataset name and a thumbnail showing the gene's expression level for each dataset sample is accessible by users. It allows for quick recognition of whether a gene shows differential expression under different experimental set-ups.

3.1.2 TACITUS

Researchers have acces to different RNA-Seq and microarrays datasets in many sizable publis repositories. Some of the most popular include NCBI GEO and Array express. These repositories contain large numbers of files requiring substantial bandwidth and specialized tools for identifying relevant subsets for research purposes. Hence, it is not easy to import or modify data from such sources. In a nutshell, however, TACITuS web-based platform provides quick query of microarray and NGS archive data. Module for handling large files and store them in a cloud and extract efficient data subset. Additionally, the technology facilitates importing of data into galaxy for further analysis. High-throughput microarray and NGS data analysis often involves processing extensive data from publicly accessible libraries. TACITuS streamlines this pre-processing task by automating several modules, enabling efficient management of large data files with agility. Additionally, it can work in a galaxy setting and has a user-friendly interface for users to analyze data [27].

TACITuS is developed using the Laravel framework and employs two databases: MariaDB for swift indexing of available datasets and MongoDB for storing both data and metadata. The data processing pipeline optimizes performance by leveraging R, C++, and PHP. The platform seamlessly integrates information from prominent sources such as NCBI GEO and Array Express with user data. TACITuS offers five essential functionalities: (i) data import, (ii) data selection, (iii) identifier mapping, (iv) data integration, and (v) Galaxy Export. A detailed, step-by-step tutorial accessible through the web interface provides in-depth insights into the implementation of these modules. The panel for "Dataset submission" within TACITuS provides a means for users to import datasets from diverse public transcriptomics resources. This includes the option to select a data source such as NCBI GEO, Array Express, or a custom source, in addition to specifying the dataset accession number and establishing whether the dataset should be classified as public or private. Following submission, the request joins a priority queue, triggering computation once resources become accessible.

Upon initial acquisition, the files are promptly downloaded, and pertinent data are meticulously archived within MongoDB databases, with the inclusion of two crucial indexes: one housing metadata attributions and the other mapping samples to their respective positions in the expression matrix through unique codes. Furthermore, a full-text index backed by Lucene is meticulously constructed for each metadata attribute, enriching the dataset's accessibility for end-users.

As for the NCBI GEO data sets, TACITuS gets the platform descriptor and makes reference links between probes and their correspondents in other platforms such as Entrez or Ensembl Gene IDs. "Selections" panel provides services like "Map Identifiers," helping in mapping probe identifiers to standard formats like Entrez (for example), which will allow integrating data among various platforms. This system uses COMBAT, Gene standardization and XPN to integrate selection items together creating one complete set of data. It entails using z-score normalization, empirical Bayesian modeling, gene normalization, and similarity analysis normalization.

Figure 2 illustrates the principle and working of the tool.

3.1.3 Array express

One of the most important online databases for functional genomics datasets is Array Express. Genome-wide gene expression data collected using microarray or next-generation sequencing (NGS) systems makes up most of the data. Array Express also offers a variety of DNA tests, including ChIP-seq and genotyping. Several assays from research are typically combined into one experiment. Depending on the sort of investigation, an assay has many definitions. An assay corresponds to one hybridization in microarray investigations (of biological sample material to an

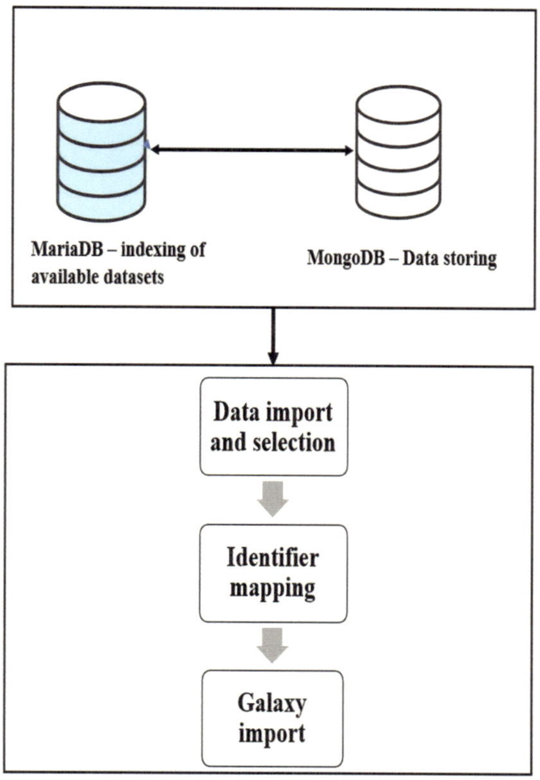

Figure 2.
Working of Tacitus: the figure explains the working of the tool, which gives a workflow of the tool.

array). A read-out (sequencing) of one library constitutes an assay for NGS investigations [28].

Array express is one of the repositories that follow the MIAME (minimal information about a microarray experiment) standards and serves as the main database of published data or those obtained through joint projects. MGED society also recommends Array Express together with GEO and CiBEX for confidential storage prior to publication. Once published, data is open source. In the last 2 years, the Array Express Repository has expanded and consists of over 50,000 hybridisations across 1650 experiments. These experiments are more than 90% for gene expression profiling, and the rests are some array-based chromatin immunoprecipitation, or comparative genomics studies. It comprises more than 200 species among which there is a considerable representation of human, mouse, Arabidopsis, yeast and rat.

The three methods that are used in data submission to the Array Express repository include; personal, internet and postal delivery. Firstly, web-based submissions are recommended for up to ~20 hybridizations, utilizing the MIA Express online tool. A batch-loader for larger experiments via this route is in testing and slated for release in 2006. Secondly, experiments of varying types and sizes can be submitted as spreadsheets, employing a template generation system for user convenience. Thirdly, laboratories with local databases can use MAGE-ML or MAGE-TAB formats for automated data export directly to Array Express. The curation process involves meticulous checking for MIAME compliance, including raw and processed data presence, accuracy and completeness of biological information, and data consistency. Array Express has been sought by journals to provide a MIAME assessment service, with legacy data set to display MIAME scores in the user interface and be accessible to reviewers supporting publications. The MIAME scoring is currently used internally for Data Warehouse selection [28].

3.2 Alignment

3.2.1 STAR

Due to the non-contiguous transcript structure, relatively small read lengths, and continuously rising throughput of the sequencing technologies, accurate alignment of high-throughput RNA-seq data is a difficult and unresolved topic. Read length limitations, high mapping error rates, slow mapping speeds, and mapping biases are all problems with the RNA-seq aligners that are currently on the market. We created the Spliced Transcripts Alignment to a Reference (STAR) software based on an undocumented RNA-seq alignment algorithm that uses sequential maximum mappable seed search in uncompressed suffix arrays followed by seed clustering and stitching procedure to align our large (>80 billion reads) ENCODE Transcriptome RNA-seq dataset. On a modest 12-core server, STAR performs better than other aligners by a factor of >50 when it comes to mapping speed, matching 550 million 2 76 bp paired-end reads to the human genome in an hour while also enhancing alignment sensitivity and precision. STAR is capable of mapping full-length RNA sequences, non-canonical splices, chimeric (fusion) transcripts, unbiased de novo identification of canonical junctions, and non-canonical splices. We experimentally confirmed 1960 unique intergenic splice junctions with an 80–90% success rate using Roche 454 sequencing of reverse transcription polymerase chain reaction amplicons, confirming the excellent accuracy of the STAR mapping technique.

While many RNA-seq aligners extend contiguous short read mappers, STAR takes a unique approach by directly aligning non-contiguous sequences to the reference genome. The STAR algorithm involves two main steps: seed searching and clustering/stitching/scoring. In the seed search phase, STAR identifies Maximal Mappable Prefixes (MMPs), akin to concepts in large-scale genome alignment tools. MMP represents the longest substring of a read sequence that exactly matches one or more substrings of the reference genome. STAR's sequential MMP search efficiently detects splice junctions, making it notably faster than comparable tools. This approach, implemented through uncompressed suffix arrays, allows for precise splice junction detection in a single pass, without prior knowledge of junction loci. The binary nature of SA search results in efficient, logarithmic scaling search times, particularly advantageous for large genomes, and facilitates accurate alignment of multimapping reads [29].

3.2.2 TOPHAT2

By sequencing transcribed RNA molecules in cells, RNA-seq allows for a more comprehensive understanding of transcription activities. In this case, analyses of RNA-seq data are done to determine the expressed genes and their amount or abundance in a cell. The first stage is mapping the RNA-seq reads back onto the reference genome and it has its own specific challenges associated with sequence alignment. The gene structures of eukaryotic genome involve intronic sequences necessitating the RNA-Seq alignment software to be able to perform gapped (spliced) alignment with varying intron lengths. In addition to this, there is a lot of processed pseudogenes in the human genome that might lead to alignment issues for read spanning exons. Mature mRNAs have an average size of about 2227 bp per transcript, with 235 bp corresponding to each exon on average, and the number of exons in a typical transcript is around 9.5 bp. The shorter the read length, the higher are the alignment complexities and about twenty percent of junction-spanning reads prove to be problematic because of very short 'anchors' that reach into the exon part of a gene. This makes it difficult to get the correct alignment especially for algorithms based on k-mers initial mapping.

In RNA-sequencing (RNA-seq) research, one popularly used spliced aligner is known as TopHat. TopHat 2 is the last one and it can align reads obtained with the most modern sequencing technologies to any reference genome allowing for variable length indel. In addition to de novo spliced alignment, TopHat version 2.0 is also capable of aligning reads across fusion breaks as they might occur after genomic translocation. TopHat2 produces sensitive and precise alignments, even for highly repetitive genomes or in the presence of pseudogenes, by combining the capacity to find novel splice sites with direct mapping to existing transcripts [30].

3.3 Assembly

3.3.1 BinPacker

Introduced by Heber, the splicing graph serves as a foundational concept in BinPacker. BinPacker works by constructing directed acyclic splicing graphs comprising of nodes, which represent exons, and edges, which represent splicing. Although nodes represent contiguous sequences in the genome that exclude alternative splicing events, they may not necessarily correspond to actual exons because of such factors like sequencing errors and poor gene expression. The tool construct splices graph with respect to an expressed gene by incorporating data from RNA-seq and it looks for the best cover with its path edges and explains each splicing graph via iteration of bin packing problems providing sufficient evidence of all the splicing events allowing whole length recovery of the transcripts. Unlike other assemblers, which sequence the transcripts using the de Bruijn graph, BinPacker sequences the spliced graph with the coverage information. According to the assumption that every splicing graph represent the single manifest transcript; every splicing graph symbolize the whole alternative splicing transcripts at each position. With the use of overlapping sequence reads and junctions, the binpacking method attempts to maximize the edge-path-cover for each splicing graph to recover the set of transcripts that may be put together [31].

3.4 Gene count generation

3.4.1 Feature counts

Next-generation sequencing technologies produce vast amounts of short reads that are typically aligned to a reference genome. One crucial aspect of downstream analysis involves determining the number of reads associated with each genomic feature, such as exons or genes. This process, known as read summarization, is essential for various genomic analyses but has not been extensively explored in the literature. A notable tool for read summarization is feature Counts, designed to count reads from both RNA and genomic DNA sequencing experiments. This program employs efficient chromosome hashing and feature blocking techniques, resulting in significantly faster performance (approximately tenfold improvement for gene-level summarization) and reduced memory requirements compared to existing methods. Feature Counts is versatile, accommodating single or paired end reads and offering a range of options tailored to different sequencing applications [32].

Feature Counts takes aligned reads in SAM or BAM format and genomic features in GFF or SAF format as input. The read input format is automatically detected, and both the read alignment and feature annotation should correspond to the same reference genome. SAM or BAM files provide detailed alignment information, including chromosome mapping and alignment specifics. Genomic features, specified in GFF or SAF format, include information like feature identifier, chromosome name, start and end positions, and strand. The tool supports strand-specific read counting when provided. It accommodates various reference sequence numbers, counting either individual features or grouping them into meta-features, such as genes. Paired or unpaired reads are supported, with paired reads counting fragments.

Feature Counts ensures precise read assignment by evaluating the mapping location of each base in a read or fragment against genomic features, accounting for gaps like insertions, deletions, and exon–exon junctions. A hit is registered with any overlap of 1 bp or more between the read/fragment and a feature. Meta-features receive hits if overlapping with any component feature. Multi-overlap reads, overlapping multiple features or meta-features, can be excluded or counted based on experiment type. For RNA-seq, excluding multi-overlap reads is suggested, while counting them is recommended for ChIP-seq, considering potential regulatory effects on overlapping genes. Chromosome hashing facilitates quick matching of reference sequence names for efficient analysis [32].

3.5 Differential gene expression

3.5.1 NOISeq

A non-parametric method for analyzing the differential expression of RNseq data is termed NOISeq. By contrasting the number of reads for each gene in samples taken under the same condition, NOISeq generates a null or noise distribution of count changes. The change in count number between two conditions for a certain gene is then evaluated using this reference distribution to determine whether it is most likely noise or a genuine differential expression. The method is implemented in two different ways: NOISeq-real uses replicates to compute the noise distribution when they are available, while NOISeq-sim simulates them when replication is not possible [33].

3.6 Automated pipelines

3.6.1 TALON

It is well known that alternative splicing regulates gene expression and plays a significant role in both healthy development and disease states. Despite increasingly powerful computational techniques, short-read RNA-seq is unable to determine full-length transcript isoforms even while it is accurate and cost-effective for quantification. Platforms for long-read sequencing, such those from Pacific Biosciences (PacBio) and Oxford Nanopore (ONT), avoid the difficulties of short-read transcript reconstruction. The ENCODE4 pipeline for platform-independent analysis of long-read transcriptomes, TALON, is introduced here. For both straightforward investigations and bigger initiatives, TALON can track both known and innovative transcript models as well as their expression levels across datasets. With the help of these characteristics, TALON users will be able to overcome the limits of short-read data and carry out isoform detection and quantification on current and upcoming long-read platforms uniformly [34].

3.6.2 TCC-GUI

Differential expression (DE) analysis of RNA-Seq count data is a critical stage in the process. For this reason, we already created the TCC R/Bioconductor package. Although this package has the distinctive ability to include a reliable normalization mechanism, only R users have been able to utilize it thus far. Therefore, for non-R users, there is a need for a DE analysis alternative to TCC. The TCC-GUI is developed in R and packaged as a Shiny application. It includes all of the key TCC features, such as strong normalization for DE pipelines and the creation of simulation data under varied scenarios. Additionally, it includes I tools for exploratory analysis, such as the average silhouette score, (ii) visualization tools like the volcano plot and heatmap with hierarchical clustering, and (iii) a reporting tool that uses R Markdown [35]. **Table 2** gives a comprehensive information of traditional and current tools its run time along with its memory usage.

4. Concluding remarks

An essential technique for functional genomics and related fields today is computational transcriptomics. The success of the mature discipline of computational transcriptomics, which uses high throughput methods like cDNA microarray and RNA sequencing, is significantly impacted by these methods. The field of bioinformatics offers a variety of databases, programmes, and automated pipelines for processing and statistical analysis of high throughput data. In order to evaluate the biological meaning from experimental data, this chapter presents an overview of key

Tool usage	Tool name	Key feature	Run time	Memory usage (RAM space)	Input	Output
Alignment	HISAT2	Provides greater alignment accuracy for reads with SNPs	8.28 ms per read	5 gb	Fastq	BAM
	STAR	Different parts of reads can be aligned to different genomic position with the reference	2 hrs	27gb	fastq	BAM
	TOPHAT2	Fast splice junction mapper	40 mins	4 Gb	Fastq	BAM files
Gene count	HTSEQ	Provides the total gene counts	22.7mins	101 Mb	BAM file	Count table
	FEATURECOUNT	Provides the total gene counts	1 min	16 Mb	Bam file	Count table
Differential expression	DESEQ2	provides a collapse function technical duplicates that can help combine the counts into a single column of the count matrix.	15–20 mins	4 Gb/thread	BAM file	Expression values
	BALLGOWN	Ballgown offers a variety of quick, easy statistical techniques to determine if transcripts differ in expression depending on the conditions of an experiment or a continuous covariate.	20–25 mins	Comparatively more memory usage	BAM file	Expression values
	CUFFDIFF	It produces a number of output files that contain test results for changes in expression at the level of transcripts, primary transcripts, and genes.	30 mins	72 Gb	BAM file	Expression values
	CUFFLINK	The aligned reads from two or more conditions and reports genes and transcripts that are differentially expressed using a rigorous statistical analysis.	30 mins	72 Gb	BAM file	Expression values
	NOISEQ	Noiseq provides expressions of genes without replicates	15–20 mins	8gb	BAM file	Expression values
Denovo Assembly	TRINITY	To extract full-length splicing isoforms and separate transcripts originating from paralogous genes, Trinity divides the sequencing data into several unique de Bruijn graphs, each of which represents the transcriptional complexity at a particular gene or locus.	2 hrs/ millime bp	20 Gb	Fastq file	Contigs
	rnaSPADES	Enable to capture the transcripts and produce high quality contigs	1 hr	512 Mb/thread	Fastq file	Contigs
	BINPACKER	Able to assemble splice junctions	1 hr	Less memory usage	Fastq file	Contigs

Table 2.
Comparison of different transcriptome analysis tools.

computational components that are used for RNA-seq data processing. The Chapter also gives a thorough overview of the tools and databases that are often utilized in transcriptomics research. It can conclude that the recent advancements in transcriptomics analysis have a new boom in the next generation sequencing field. This has made the researchers explore more on the Transcriptomics and its applications in solving complex biological problems.

Conflict of interest

The authors declared no potential conflicts of interest with respect to the research, authorship, and/or publication of this article.

Author details

Vidya Niranjan[1]*, Lavanya Chandramouli[1], Pooja SureshKumar[1] and Jitendra Kumar[2]*

1 Department of Biotechnology, R.V. College of Engineering, Bengaluru, Karnataka, India

2 Biotechnology Industry Research Assistance Council (BIRAC), New Delhi, India

*Address all correspondence to: vidya.n@rvce.edu.in and md.birac@nic.in

References

[1] Cobb M. 60 years ago, Francis crick changed the logic of biology. PLoS Biology. 2017;**15**(9):e2003243. DOI: 10.1371/journal.pbio.2003243

[2] Mattick JS. Challenging the dogma: The hidden layer of non-protein-coding RNAs in complex organisms. BioEssays. 2003;**25**(10):930-939

[3] Evans JS. Principles of molecular biology and biomacromolecular chemistry. Reviews in Mineralogy and Geochemistry. 2003;**54**(1):31-56

[4] Ratti M, Lampis A, Ghidini M, Salati M, Mirchev MB, Valeri N, et al. MicroRNAs (miRNAs) and long non-coding RNAs (lncRNAs) as new tools for cancer therapy: First steps from bench to bedside. Targeted Oncology. 2020;**15**(3):261-278. DOI: 10.1007/s11523-020-00717-x

[5] Amaro A, Petretto A, Angelini G, Pfeffer U. Advancements in omics sciences. In: Translational Medicine. Cambridge, Massachusetts, United States: Academic Press; 2016. pp. 67-108

[6] Ramsköld D, Wang ET, Burge CB, Sandberg R. An abundance of ubiquitously expressed genes revealed by tissue transcriptome sequence data. PLoS Computational Biology. 2009;**5**(12):e1000598

[7] Lavanya C, Upadhyaya A, Neogi AG, Niranjan V. Identification of novel regulatory pathways across normal human bronchial epithelial cell lines (NHBEs) and peripheral blood mononuclear cell lines (PBMCs) in COVID-19 patients using transcriptome analysis. Informatics in Medicine Unlocked. 2022;**31**:100979

[8] Lowe R, Shirley N, Bleackley M, Dolan S, Shafee T. Transcriptomics technologies. PLoS Computational Biology. 2017;**13**(5):e1005457

[9] Wang Z, Gerstein M, Snyder M. RNA-Seq: A revolutionary tool for transcriptomics. Nature Reviews. Genetics. 2009;**10**(1):57-63. DOI: 10.1038/nrg2484

[10] Patel RK, Jain M. NGS QC toolkit: A toolkit for quality control of next generation sequencing data. PLoS One. 2012;**7**(2):e30619

[11] Leggett RM, Ramirez-Gonzalez RH, Clavijo BJ, Waite D, Davey RP. Sequencing quality assessment tools to enable data-driven informatics for high throughput genomics. Frontiers in Genetics. 2013;**4**:288. DOI: 10.3389/fgene.2013.00288

[12] Bolger AM, Lohse M, Usadel B. Trimmomatic: a flexible trimmer for Illumina sequence data. Bioinformatics. 2014;**30**(15):2114-2120

[13] Liu X, Yan Z, Wu C, Yang Y, Li X, Zhang G. FastProNGS: Fast preprocessing of next-generation sequencing reads. BMC Bioinformatics. 2019;**20**(1):345. DOI: 10.1186/s12859-019-2936-9

[14] Schatz MC, Delcher AL, Salzberg SL. Assembly of large genomes using second-generation sequencing. Genome Research. 2010;**20**(9):1165-1173. DOI: 10.1101/gr.101360.109

[15] Bushmanova E, Antipov D, Lapidus A, Prjibelski AD. rnaSPAdes: A de novo transcriptome assembler and its application to RNA-Seq data. GigaScience. 2019;**8**(9):giz100

[16] Grabherr MG, Haas BJ, Yassour M, Levin JZ, Thompson DA, Amit I, et al. Full-length transcriptome assembly from

RNA-Seq data without a reference genome. Nature Biotechnology. 2011;**29**(7):644-652. DOI: 10.1038/nbt.1883

[17] Conesa A, Madrigal P, Tarazona S, Gomez-Cabrero D, Cervera A, McPherson A, et al. A survey of best practices for RNA-seq data analysis. Genome Biology. 2016;**17**(1):1-19

[18] Kim D, Langmead B, Salzberg SL. HISAT: A fast spliced aligner with low memory requirements. Nature Methods. 2015;**12**(4):357-360. DOI: 10.1038/nmeth.3317

[19] Kukurba KR, Montgomery SB. RNA sequencing and analysis. Cold Spring Harbor Protocols. 2015;**2015**(11):951-969. DOI: 10.1101/pdb.top084970

[20] Abbas-Aghababazadeh F, Li Q, Fridley BL. Comparison of normalization approaches for gene expression studies completed with high-throughput sequencing. PLoS One. 2018;**13**(10):e0206312. DOI: 10.1371/journal.pone.0206312

[21] Zhao S, Ye Z, Stanton R. Misuse of RPKM or TPM normalization when comparing across samples and sequencing protocols. RNA. 2020;**26**(8):903-909. DOI: 10.1261/rna.074922.120

[22] Love MI, Huber W, Anders S. Moderated estimation of fold change and dispersion for RNA-seq data with DESeq2. Genome Biology. 2014;**15**(12):1-21

[23] Frazee AC, Pertea G, Jaffe AE, Langmead B, Salzberg SL, Leek JT. Ballgown bridges the gap between transcriptome assembly and expression analysis. Nature Biotechnology. 2015;**33**(3):243-246. DOI: 10.1038/nbt.3172

[24] Trapnell C, Roberts A, Goff L, Pertea G, Kim D, Kelley DR, et al. Differential gene and transcript expression analysis of RNA-seq experiments with TopHat and cufflinks. Nature Protocols. 2012;**7**(3):562-578. DOI: 10.1038/nprot.2012.016

[25] Li W, Jiang T. Transcriptome assembly and isoform expression level estimation from biased RNA-Seq reads. Bioinformatics. 2012;**28**(22):2914-2921

[26] Barrett T, Suzek TO, Troup DB, Wilhite SE, Ngau WC, Ledoux P, et al. NCBI GEO: Mining millions of expression profiles—Database and tools. Nucleic Acids Research. 2005;**33**(suppl_1):D562-D566

[27] Alaimo S, Di Maria A, Shasha D, Ferro A, Pulvirenti A. TACITuS: Transcriptomic data collector, integrator, and selector on big data platform. BMC Bioinformatics. 2019;**20**(9):1-11

[28] Athar A, Füllgrabe A, George N, Iqbal H, Huerta L, Ali A, et al. ArrayExpress update–from bulk to single-cell expression data. Nucleic Acids Research. 2019;**47**(D1):D711-D715

[29] Dobin A, Davis CA, Schlesinger F, Drenkow J, Zaleski C, Jha S, et al. STAR: ultrafast universal RNA-seq aligner. Bioinformatics (Oxford, England). 2013;**29**(1):15-21. DOI: 10.1093/bioinformatics/bts635

[30] Kim D, Pertea G, Trapnell C, Pimentel H, Kelley R, Salzberg SL. TopHat2: Accurate alignment of transcriptomes in the presence of insertions, deletions and gene fusions. Genome Biology. 2013;**14**(4):1-13

[31] Liu J, Li G, Chang Z, Yu T, Liu B, McMullen R, et al. BinPacker: Packing-based De novo transcriptome assembly from RNA-seq data. PLoS Computational Biology. 2016;**12**(2):e1004772. DOI: 10.1371/journal.pcbi.1004772

[32] Liao Y, Smyth GK, Shi W. featureCounts: An efficient general

purpose program for assigning sequence reads to genomic features. Bioinformatics. 2014;**30**(7):923-930

[33] Tarazona S, Furió-Tarí P, Turrà D, Pietro AD, Nueda MJ, Ferrer A, et al. Data quality aware analysis of differential expression in RNA-seq with NOISeq R/bioc package. Nucleic Acids Research. 2015;**43**(21):e140-e140

[34] Wyman D, Balderrama-Gutierrez G, Reese F, Jiang S, Rahmanian S, Forner S, et al. A technology-agnostic long-read analysis pipeline for transcriptome discovery and quantification. Biorxiv. 2019;**672931**

[35] Su W, Sun J, Shimizu K, Kadota K. TCC-GUI: A shiny-based application for differential expression analysis of RNA-Seq count data. BMC Research Notes. 2019;**12**(1):1-6

Chapter 3

Approaches to Improve the Immunogenicity of Plasmid DNA-Based Vaccines against COVID-19

Mariya Borisovna Borgoyakova,
Ekaterina Aleksandrovna Volosnikova,
Aleksander Alekseevich Ilyichev and Larisa Ivanovna Karpenko

Abstract

Plasmid DNA-based vaccines are attracting considerable interest because of their potential as a platform technology that can be used for a variety of purposes from prevention to therapy. The COVID-19 pandemic has stimulated the development of this platform. The DNA vaccine against COVID-19, developed by Zydus Cadila, was the world's first DNA vaccine approved for human vaccination. However, the problem of low immunogenicity of DNA vaccines has not yet been completely solved. This article will describe the authors' experience in creating plasmid DNA-based vaccines against COVID-19, including the design of target antigens, artificial polyepitope T-cell immunogens, delivery of the resulting plasmid constructs using polycationic biodegradable polymers, and producing artificial self-assembled particles incorporating the recombinant protein and DNA vaccine.

Keywords: plasmid DNA-based vaccine, SARS-CoV-2, immunogenicity, DNA-vaccine delivery, polycationic biodegradable polymers, self-assembled particles

1. Introduction

Plasmid DNA vaccines have attracted considerable interest as a platform technology that can be used for a variety of purposes from prevention to therapy. DNA vaccines are advantageous because, like recombinant viral vectors and mRNA vaccines, they induce T-cell and humoral immunity with a good safety profile. The immune response induced by the DNA vaccine is antigen-specific and is not produced to the construction itself, which allows its repeated administration, including prime-boost regimens. DNA vaccines are not pathogenic agents; hence, there is no risk of infection when administered, and numerous clinical studies have shown their high safety [1–4]. An important advantage is also the stability

IntechOpen

of DNA in a wide temperature range and, hence, the lack of requirements for cold chain maintaining during storage and transportation of DNA vaccines [5, 6]. Production of plasmid DNA is much faster and easier than the production of live, inactivated, or subunit vaccines; it is much safer, especially compared to vaccines based on live or inactivated pathogens. When new challenges arise, DNA vaccines can be easily adapted to new targets, and they are stable at room temperature, which positively distinguishes them from mRNA vaccines, requiring storage at low temperatures [7].

The COVID-19 pandemic stimulated the development of this platform. The ZyCoV-D DNA vaccine against COVID-19, developed by Cadila Healthcare, was the world's first DNA vaccine approved for human vaccination [8], showing 66.6% efficacy against symptomatic COVID-19 and 100% efficacy against the development of severe disease in phase III clinical trials [9].

Although DNA vaccines have demonstrated the ability to induce humoral and cellular immunity in small laboratory animal models, intramuscular administration of naked DNA to humans involves increasing the DNA dose to several mg to achieve high efficacy. Low immunogenicity, when administered with naked form, is the main disadvantage of DNA vaccines. This is due to the fact that DNA delivery into cells is difficult because of the large size of the plasmids, their hydrophilic nature, and the overall negative charge of the nucleic acid. In addition, DNA entering the intercellular space can be degraded by nucleases. All this reduces the immunogenicity of plasmids. A large number of studies have focused on developing strategies to improve the immunogenicity of DNA vaccines.

One strategy is the optimization of the coding sequence itself: for example, codon-optimization has become a routine procedure for creating target genes in vaccine plasmids, for which a large number of tools have been created [10, 11]. Another strategy involves incorporating different sequences into the target genes: various regulatory sequences (to ensure high expression of target genes, increase stability of mRNA molecules, and enhance translation [12–15]), leader sequences (for efficient secretion of a protein product from the cell [10, 16]), signal sequences, such as ubiquitin and LAMP (to enable protein processing in the cell and increase the efficiency of antigen presentation by MHC I and MHC II molecules [17, 18]).

The low immunogenicity of DNA vaccines is largely associated with low transfection activity; therefore, various methods are being actively developed to increase the efficiency of the penetration of injected DNA into cells. This has led to the creation of a wide range of physical and chemical methods of DNA-vaccine delivery [19]. Through the analysis of the publicly available data on clinical studies of DNA vaccines against COVID-19, we can see that most researchers inject plasmids using special devices: electroporators or jet injectors [8, 20–24]. An interesting trend is the use of a live bacterial vector, such as lactic acid bacteria or Salmonella, which provides not only a strong systemic immune response but also a mucosal one [25, 26]. Being noninvasive, this method is preferred for most patients because it is not associated with pain [27, 28].

Another trend is the development of delivery vehicles based on cationic carriers. The most common approach to packaging DNA vaccines is using of cationic liposomes, which increase the efficiency of cell transfection during intramuscular injection but have serious drawbacks such as toxicity due to the ability to destabilize the cell membrane [29]. Various cationic polymers of natural (chitosan, gamma-polyglutamic acid, and hyaluronic acid) and synthetic origin (polylactideglycolide, polyethylenimine, and others) are alternatives to lipids [30].

This article describes the authors' original approaches to constructing plasmid DNA vaccines against COVID-19, including the development of target antigens and artificial polyepitope T-cell immunogens, delivery of the resulting plasmid constructs using polycationic biodegradable polymers, and production of artificial self-assembled particles, including the recombinant protein and DNA vaccine.

2. Methodology

The sequence encoding the receptor-binding domain (RBD) of SARS-CoV-2 spike protein published in GenBank (320 V-542 N, MN908947) was used to design the RBD immunogen. The sequence encoding leader peptide MMRTLILAVLLVYFCATVHC was added to the RBD sequence using standard gene engineering methods. Selection of preferred codons, that is, avoiding rare codons with low utilization, and optimizing the secondary RNA structure for efficient expression in mammalian cells have been performed using the GeneOptimizer software [31]. The resulting nucleotide sequence was synthesized and cloned into the pVAX1 vector (DNA-Syntez LLC, Moscow, Russia) under the early promoter of human cytomegalovirus (CMV). After confirming the structure of the construct by Sanger sequencing, it was designated as pVAXrbd.

A strategy based on combining conserved fragments from the main viral proteins S, N, M, and E containing clusters of overlapping epitopes into a single molecule was used to design the T-cell immunogen. T-cell epitope prediction was performed using NetMHCpan-4.1 software and the Immune Epitope Database 2.22, IEDB 2.22 (**Tables 1** and **2**). To study the DNA vaccine in an animal model, additional mouse MHC I and II restriction epitopes were included in the immunogen sequence. After combining the fragments into one sequence, the universal T-helper epitope PADRE was added to its N-terminus and the marker epitope EPFRDYVDRFYKTLR recognized by 29F2 monoclonal antibodies was added to the C-terminus. The sequence encoding ubiquitin was added to the artificial immunogen sequence, and its whole sequence was optimized taking into account the codons that ensure efficient expression in mammalian cells using the Jcat program [32]. The resulting gene was

MHC II Allele	Peptide	Adjusted Rank
H2-Ied	DDQIGYYRRATRRIR	0.11
H2-Ied	NYNYLYRLFRKSNLK	0.17
H2-Ied	VKPSFYVYSRVKNLN	0.33
H2-Iad	EMIAQYTSALLAGTI	2.80
H2-Ied	ASAFFGMSRIGMEVT	4.35
H2-Ied	QYIKWPWYIWLGFIA	7.05
H2-Ied	CFVLAAVYRINWITG	7.30
H2-IEd	GTWLTYTGAIKLDDK	10.15

The peptides with an adjusted rank value of 10.5 or less are listed in the table.

Table 1.
Th-epitopes from SARS-CoV-2 proteins restricted by various BALB/c mouse MHC II molecules (the lower the value in the rightmost column, the stronger the binding between the MHC II molecule and the peptide) were predicted using the IEDB 2.22 tool.

MHC I Allele	Peptide	Score
H-2-Kd	AYSNNSIAI	0.837637
H-2-Kd	QYIKWPWYI	0.738706
H-2-Kd	YYRRATRRI	0.414253
H-2-Ld	LPPLLTDEM	0.384921
H-2-Dd	SAPHGVVFL	0.370438
H-2-Ld	WPWYIWLGF	0.355595
H-2-Kd	NWITGGIAI	0.347568
H-2-Ld	TPSGTWLTY	0.32614
H-2-Ld	FPQSAPHGV	0.282244
H-2-Dd	FAPSASAFF	0.242852
H-2-Kd	GFIAGLIAI	0.237561
H-2-Kd	QFAPSASAF	0.218262
H-2-Kd	KHIDAYKTF	0.172696
H-2-Ld	FPRGQGVPI	0.160547
H-2-Ld	NSIAIPTNF	0.148747
H-2-Kd	TWLTYTGAI	0.144436
H-2-Ld	SPDDQIGYY	0.126441
H-2-Ld, H-2-Kd	IAIPTNFTI	0.119486
H-2-Ld	QSAPHGVVF	0.159378
H-2-Kd	NFKDQVILL	0.113114
H-2-Ld	YNYKLPDDF	0.101053

Peptides with a score value above 0.1 are listed in the table.

Table 2.
CTL-epitopes from SARS-CoV-2 proteins that are restricted by different BALB/c mouse MHC I molecules (the higher the value in the rightmost column, the stronger the binding between the MHC I molecule and the peptide) predicted with the NetMHCpan-4.1 tool.

synthesized and cloned as part of the pVAX1 vector plasmid. The corresponding plasmid was named pBSI-COV-Ub.

To evaluate target gene expression, HEK 293-T cells were transfected with plasmids encoding RBD and BSI-COV-Ub proteins. Gene expression was determined by assessing mRNA level and protein content using RT-PCR and Western blot analysis, respectively.

RBD protein was conjugated to polyglucin-spermidine, which is a polycationic carrier used to deliver DNA vaccines [33–36]. PGS and PGS-RBD conjugates were used to generate two types of particles (**Figures 1** and **2**). For the formation of CCV-RBD particles, DNA was mixed with the PGS-RBD conjugate at a ratio of 50 μg of protein for every 100 μg of DNA, incubated for 5 min at room temperature, then PGS was added in 10-fold excess and incubated for another 1 h at 4°C. To form pVAXrbd-PGS particles, DNA was mixed with PGS conjugate in the mass ratio 1:10 and incubated for 1 h at 4°C. The efficiency of complex formation was assessed by changes in electrophoretic mobility of DNA in 1% agarose gel, as well as by gel-filtration on a

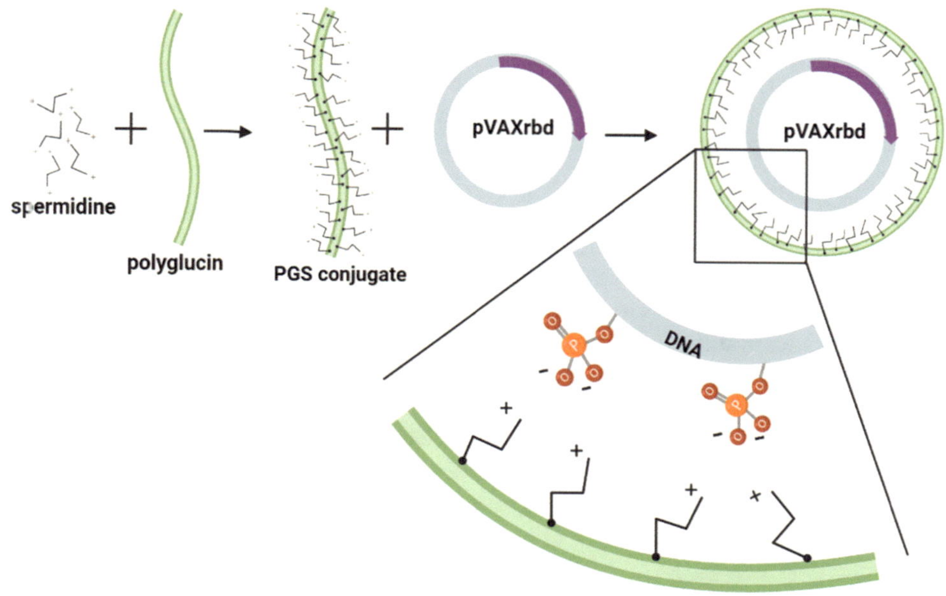

Figure 1.
Schematic representation of the pVAXrbd-PGS particle assembly and interaction between polycationic polymer PGS and DNA.

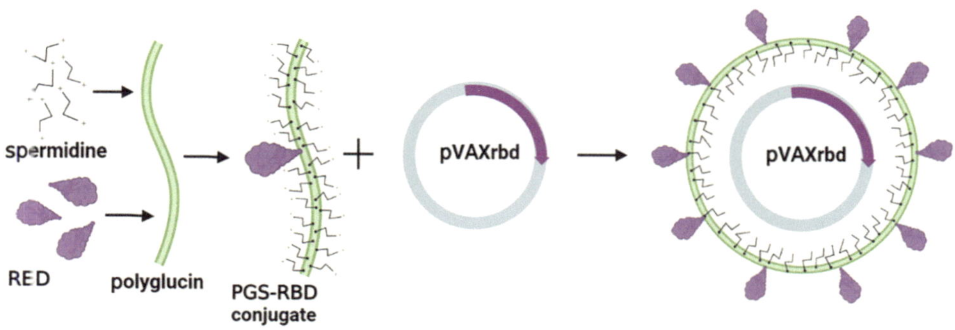

Figure 2.
Schematic representation of the PGS-RBD conjugation and assembly of the combined CCV-RBD vaccine particle.

Sepharose CL-6B column [34]. Samples for gel filtration were applied to the column in equimolar amounts of nucleotide material.

To evaluate the immunogenicity of the experimental vaccine constructs, BALB/c mice were twice or thrice immunized intramuscularly or intramuscularly in combination with electroporation. All experiments involving animals were carried out in compliance with the principles of humanity according to the protocols approved by the Bioethical Committee of the SRC VB Vector. Electroporation was performed using a CUY21 EDIT II electroporator and LF650P5 5-mm forceps electrode (BEX CO, Ltd.). The following electroporation protocol was used: rectangular direct and reverse polarity direct current with 3 pulses at 12 V for 30 and 950 ms interval with a current limit of 45 mA [37]. Ten to fourteen days after the second/third immunization, the mice were taken out from the experiment by cervical dislocation, sera were collected from the blood samples, and the spleens were isolated to study the immune response.

Preparation of the material for immunological studies (ELISA, VN, ELISpot, ICS) was performed as described previously [34].

ELISA was carried out using recombinant RBD protein sorbed onto 96-well plates in 1 mkg per ml concentration. Viral neutralizing activity of the obtained sera was assessed in an *in vitro* neutralization assay using live nCoV/Victoria/1/2020 strain of the SARS-CoV-2.

The T-cell immune response was evaluated using IFN-γ-ELISpot technology corresponding manufacturers' protocol. Splenocytes were stimulated with a pool of peptides from the RBD protein or artificial T-cell immunogen. The ICS procedure was carried out as described previously [34].

The results were statistically processed using nonparametric Mann-Whitney analysis in GraphPad Prism 6.0 program (GraphPad Software) and presented as group mean values, mean with standard deviation (for graphic representation of ELISA, VN, and ELISpot), mean with total variation (for graphic representation of ICS), and medians with 95% confidence interval (for graphic presentation of protectivity). The differences were considered statistically significant at $p < 0.05$.

3. Results and discussion

3.1 Design of the immunogen

3.1.1 Design of an immunogen capable of inducing a humoral and cellular response

For the design of the first immunogen, the sequence encoding the receptor-binding domain (RBD) of SARS-CoV-2 spike protein published in GenBank (320 V-542 N, MN908947) was used. For SARS-CoV-2, the RBD protein is a good antigen candidate because this relatively small protein is the primary target of neutralizing antibodies [38–44].

In general, DNA vaccines can induce strong cell-mediated immunity and relatively weak humoral immunity. This is due to the fact that antigen synthesis occurs inside the cell, where it is processed and presented by MHC molecules. To enhance the humoral response, it was decided to increase antigen secretion from the cell. The original signal sequence MMRTLILAVLLVYFCATVHC, a hybrid of leader sequences of two actively secreted proteins: luciferase and fibroin (the leader peptide was developed by Shcherbakov [33]), was introduced to provide efficient transport of RBD protein out of the cell. Obtained construction was confirmed and designated as pVAXrbd.

3.1.2 Design of an artificial polyepitope T-cell immunogen

An effective vaccine against coronavirus must induce a potent T-cell response in addition to neutralizing antibodies. Virus-specific T-cell immunity (against S, N, and M proteins) was detected in people who had contracted the related SARS-CoV more than 10 years after the disease [45]. Using only the RBD as part of the DNA vaccine cannot ensure a long-term effective cellular response. Therefore, the strategy of using also the sequences that are parts of the minor virus proteins as T-cell immunogens seems to be more promising [46]. In addition, these proteins are least susceptible to variability due to less selection pressure.

The peptides listed in **Tables 1** and **2** were analyzed for conservativity for different variants of SARS-CoV-2, particularly Wuhan, Gamma, Delta, and Omicron. For each peptide, more than 87% homology to the proteins of the indicated SARS-CoV-2 strains was established.

It is worth noting that the fragments selected for inclusion in the final construct also contained epitopes restricted by a broad spectrum of HLA I (**Figure 3**) and HLA II, which confirmed immunodominance in the study of SARS-CoV-2-specific cellular immunity in humans and were conserved for different variants of the virus [47–49]. After combining the fragments into one sequence, the universal T-helper epitope PADRE was added to its N-terminus, and the marker epitope EPFRDYVDRFYKTLR recognized by 29F2 monoclonal antibodies was added to the C-terminus (**Figure 4**).

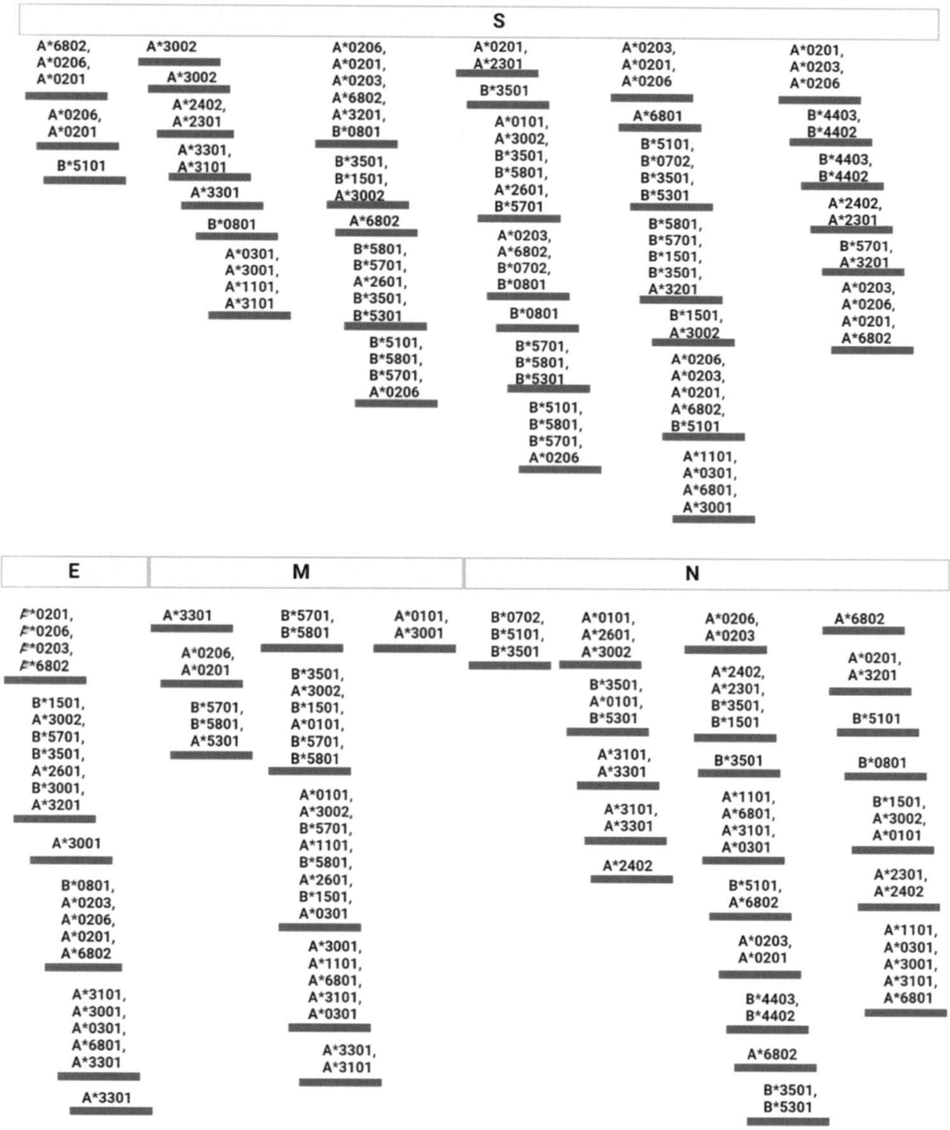

Figure 3.
Schematic arrangement of CTL epitopes predicted using the NetMHCpan-4.1 tool from SARS-CoV-2 proteins restricted by different HLA I molecules. Epitopes with a score value greater than 0.3 were taken for the schematic.

Figure 4.
General schematic structure of the polyepitope immunogen BSI-COV-Ub.

In addition, it was decided to attach ubiquitin to the N-terminus of the immunogen. The attachment of ubiquitin to the protein immunogen promotes its targeting to the proteasome, which leads to efficient processing, the release of peptides-epitopes that are presented by MHC I on the surface of the antigen-presenting cell and promote cytotoxic T lymphocyte activation [18, 50, 51], as well as activate T-helper cells as a result of cross-presentation.

The nucleotide sequence corresponding to the designed amino acid sequence was named BSI-COV-Ub (BSI is for Bazhan Sergey Ivanovich, deceased creator of the artificial immunogen) and the corresponding plasmid was named pBSI-COV-Ub.

3.2 The assessment of DNA vaccines' biological activity

Once the immunogens had been constructed and the plasmids encoding RBD and BSI-COV-Ub had been obtained, their ability to ensure synthesis of the target immunogens in eukaryotic cells needed to be assessed. It was shown that the sizes of the amplified fragments detected using RT-PCR correspond to the theoretically calculated lengths of the target genes: 750 bps for RBD and 1500 bps for BSI-COV.

Western blot assay demonstrated that RBD was detected in both cell lysate and culture medium, indicating a high degree of protein production and secretion provided by the inclusion of a special leader sequence [33]. The polyepitope BSI-COV-Ub protein was detected only in the cell lysate as a set of discrete proteins, indicating the effective processing of T-cell immunogen in the cell, which is ensured, inter alia, by the attachment of ubiquitin to it [52].

3.2.1 Immunogenicity of the DNA vaccines

To evaluate the immunogenicity of the DNA vaccine constructs, BALB/c mice were immunized with pVAXrbd and pBSI-COV-Ub. Fourteen days after a double intramuscular injection of the vaccines (100 μg/dose), blood and spleens were collected from the animals for analysis. According to ELISA results, the average titers of specific antibodies in animals immunized with pVAXrbd were approximately 1:2300, which is pretty well for a naked DNA vaccine. The immunogen encoded by pBSI-COV-Ub did not contain B-cell epitopes, so it induced no virus-specific antibodies as expected.

IFN-γ-ELISpot data showed that the group received 100 μg pVAXrbd had an average number of IFN-γ-secreting T lymphocytes (SFUs) per 10^6 splenocytes of 238, while the group received 100 μg pBSI-COV-Ub had 122 SFUs. Intracellular cytokine staining (ICS) showed a significant level of IL-2-producing CD4+ and CD8+ T lymphocytes, indicating the formation of a strong virus-specific T-cell response.

Thus, the pVAXrbd DNA vaccine construct, which encodes the RBD gene with a secretory domain, provides induction of the humoral immune response due to the efficient secretion of RBD protein from the cell. Both pVAXrbd and pBSI-COV-Ub DNA vaccines induce a virus-specific T-cell response. It should be noted, in a mouse model, it is difficult to fully assess the potential of the T-cell response of pBSI-COV-Ub because this plasmid encodes epitopes that are restricted by both mouse and human MHC.

The protective efficacy of the designed DNA vaccines was evaluated in the model of mice challenged with the SARS-CoV-2 gamma variant (strain hCoV-19/Russia/SA-17620-080521/2021). Infection was performed 2 weeks after the second immunization with an intranasal dose of 50 ID_{50}. As shown in **Figure 5**, minimal viral load was detected in mice of the group immunized with pBSI-COV-Ub. The decrease in viral RNA amount compared to the control was 3.45 lg (32.48 Ct vs. 21.00 Ct ($p < 0.01$)). The lungs of mice immunized with pVAXrbd showed a close value of 3.18 lg (31.53 Ct vs. 21.00 Ct ($p < 0.05$)) decrease in the amount of viral RNA compared to control.

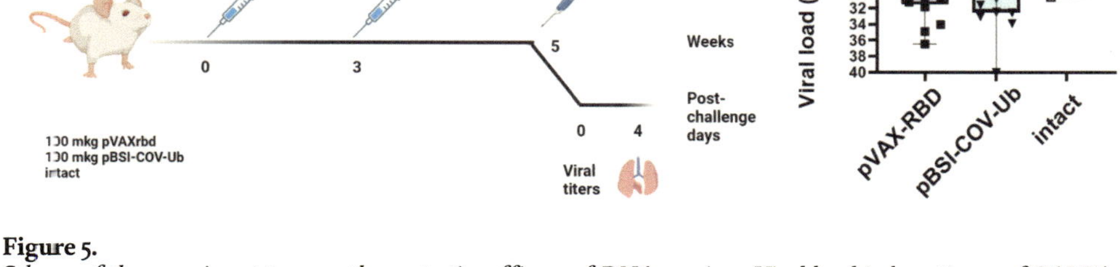

Figure 5.
Scheme of the experiment to assess the protective efficacy of DNA vaccines. Viral load in lung tissues of BALB/c mice on day 4 after challenge with SARS-CoV-2 gamma variant. Individual Ct values (threshold virus detection cycle) in PCR-RT are represented by dots; medians in the groups by histogram vertices, with numerical Ct values indicated at the base of the histograms; 95% confidence interval limits are indicated by vertical lines. We express our deep gratitude to Pyankov O.V. and his team for the challenge performing.

Thus, both immunogens in DNA format have the ability to induce protective immunity in mice against gamma variant of the SARS-CoV-2, which confirms the importance of further work with these vaccines.

3.3 DNA-vaccine delivery

3.3.1 Delivery of the DNA vaccine using polyglucin-spermidine conjugate

The use of cationic biopolymers for plasmid delivery is one of the techniques that is used to increase the efficiency of DNA vaccines [30]. The most widely used strategy is the association of DNA and the biopolymer through electrostatic interaction between the negatively charged DNA phosphate groups and the positive charges of the carrier material.

The original cationic polymer used in this work was a polyglucin-spermidine conjugate (PGS) [35]. It is important that the components of the resulting polymer are biodegradable and safe for humans, animals, and the environment; polyglucin is a glucose polymer with a molecular weight of 40,000. It is a licensed plasma substitute and is nontoxic. Spermidine is a polyamine that is found in the cells of almost all living organisms and is involved in many biological processes. The low cost, safety, and ability to lyophilize with long-term storage at 4°C provide additional technological advantages in the production and transportation of vaccine preparations containing PGS [53].

The polyglucin-spermidine conjugate can be used both for the delivery of proteins, including the recombinant RBD protein [54] and nucleic acids. Previously, polyglucin-spermidine has proven to be an effective delivery vehicle for experimental DNA vaccines against Ebola and HIV-1 [35, 36, 55]. As a component of anti-HIV vaccine, PGS passed the first phase of clinical trials, demonstrating a high safety level [56].

The optimum ratio of the components for the formation of DNA-polymer complexes was selected based on the analysis of the complexation degree obtained earlier [35]. The ratio of DNA and PGS weights to form complexes was 1:10 (**Figure 1**). The particles' assembling was assessed by a shift in electrophoretic mobility in agarose gel: the encapsulated plasmids lost their mobility in the electric field (**Figure 6A**). Electron microscopy was used to determine the size of the particles, whose values

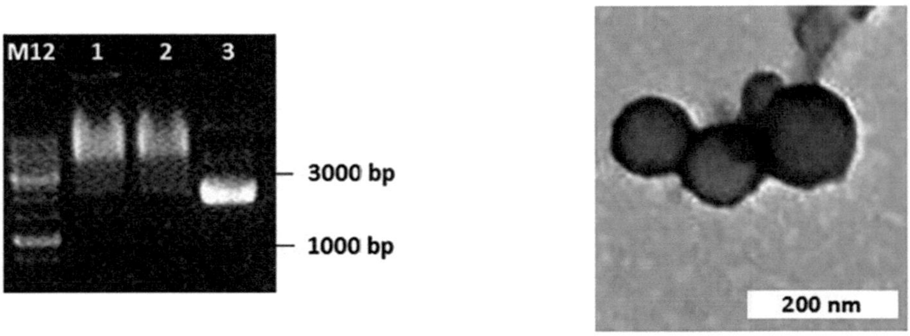

Figure 6.
Assessment of the CCV-RBD particle formation. (A) DNA encapsulation in the PGS and PGS-RBD shells by electrophoresis in a 1% agarose gel: 1, CCV-RBD; 2, pVAXrbd-PGS; and 3, naked plasmid pVAXrbd. (B) Electron micrograph of CCV-RBD particles. We express our gratitude to Zaitsev B.N. For electron microscopy results.

were in the range of 50–1000 nm. It is shown that such size of the particles is optimal for vaccine creation since they accumulate in B-cell follicles and induce a strong immune response [57].

After three-time immunizations on days 0, 14, and 28, blood was collected from the animals for analysis 6 weeks after the start of the experiment. The sera were examined for the presence of RBD-specific antibodies by ELISA and in the SARS-CoV-2 neutralize reaction. The sera of unimmunized mice were used as a control. According to ELISA results, the mean titers of specific antibodies in sera of animals immunized with pVAXrbd-PGS were approximately 40-fold higher compared with animals immunized with pVAXrbd ($p < 0.01$) and 10,000-fold higher than titers in controls ($p < 0.01$) (**Figure 7A**).

The neutralizing ability of the obtained sera was evaluated in an *in vitro* neutralization assay using live SARS-CoV-2. Sera from mice immunized with pVAXrbd-PGS neutralized the nCoV/Victoria/1/2020 SARS-CoV-2 strain at a dilution of 1:200, whereas sera from the group immunized with pVAXrbd neutralized it at a dilution of 1: 12 ($p < 0.01$). Sera from control animals showed no neutralizing activity (**Figure 7B**).

The T-cell immune response was assessed by the number of splenocytes producing IFN-γ using ELISpot technology. Splenocytes were stimulated with a pool of peptides from RBD protein. It was shown that splenocytes isolated from the spleens of mice immunized with pVAXrbd in the PGS envelope or none responded more strongly to the stimulation with viral peptides with IFN-γ release than the control group ($p < 0.01$) but without statistically significant differences between each other (**Figure 7C**).

Thus, the use of polyglucin-spermidine as a delivery vehicle enhances the immunogenicity of the DNA vaccine, including increased levels of neutralizing antibodies and T-cell response [33].

3.3.2 Combined DNA/protein particles

DNA vaccines are considered to be most effective in inducing a T-cell response and less effective in inducing a humoral response; therefore, it was decided to further

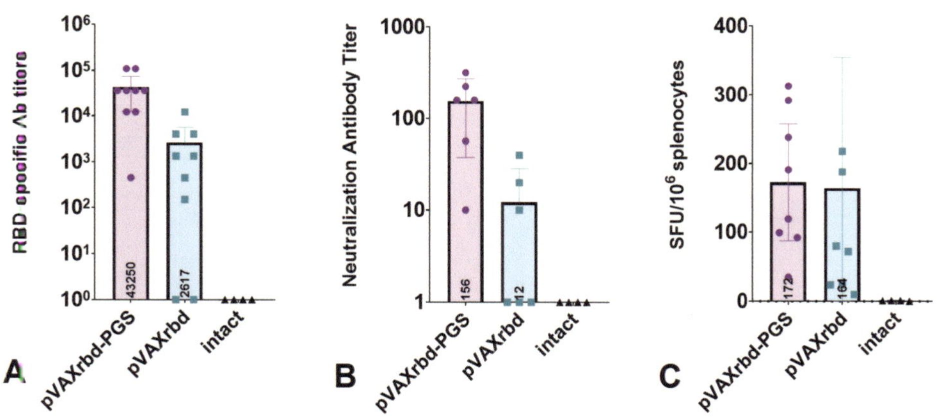

Figure 7.
Immune response induced by the pVAXrbd-PGS and pVAXrbd constructs in mice. (A) RBD-specific IgG antibody titers were assessed in ELISA. Reciprocal titer values are shown in the plot. (B) Virus neutralization activity of sera was assessed using the SARS-CoV-2 (nCoV/Victoria/1/2020 strain) neutralization reaction. Reciprocal titer values are provided in the plot. (C) T-cell response was assessed by the ELISpot method. The number of spot-forming units is shown in the plot. We express our gratitude to Pyankov O.V. and his team for the virus neutralization performing.

enhance the humoral component by incorporating the protein immunogen-RBD protein into the design [56].

For this purpose, we decided to use polyglucin-spermidine-protein RBD conjugate (PGS-RBD) for DNA vaccine packaging. The synthesis technique was proposed by Lebedev *et al.* [54, 55, 58, 59]. We previously used this approach to increase the immunogenicity of an experimental vaccine based only on the recombinant RBD protein [54]. We attached the protein to a polyglucine-spermidine conjugate and mixed it with dsRNA (as adjuvant). As a result, complexes presenting RBD on the surface and dsRNA inside were obtained due to self-assembly of conjugate and RNA. Such complexes induced virus neutralizing antibody level against SARS-COV-2 much more effectively than RBD with AL(OH)3 or RBD without adjuvant [54].

In this work, we used the PGS-RBD conjugate to deliver the pVAXrbd DNA vaccine. CCV-RBD particles (CCV-RBD stands for CombiCoronaVac-RBD) were obtained by mixing PGS-RBD with plasmid pVAXrbd as described earlier [34]. The assembling of the complexes was confirmed by a loss of the encapsulated plasmids' mobility in the electric field (**Figure 6A**). Preservation of plasmid DNA structure in CCV-RBD and pVAXrbd-PGS complexes was conserved by UV spectroscopy; the spectra of these preparations had a characteristic DNA peak at 260 nm. Using electron microscopy, the particle sizes appeared to be in the range of 50–1000 nm (**Figure 6B**) are comparable to the particles formed by pVAXrbd and PGS without protein [34].

To assess changes in the binding kinetics of RBD after including in CCV-RBD particles with the known mAb iB14, a study was carried out using biolayer interferometry performed on the Octet K2 device (Pall Fortebio, USA). The dissociation constants of RBD and CCV-RBD particles with iB14 showed close values, suggesting protein

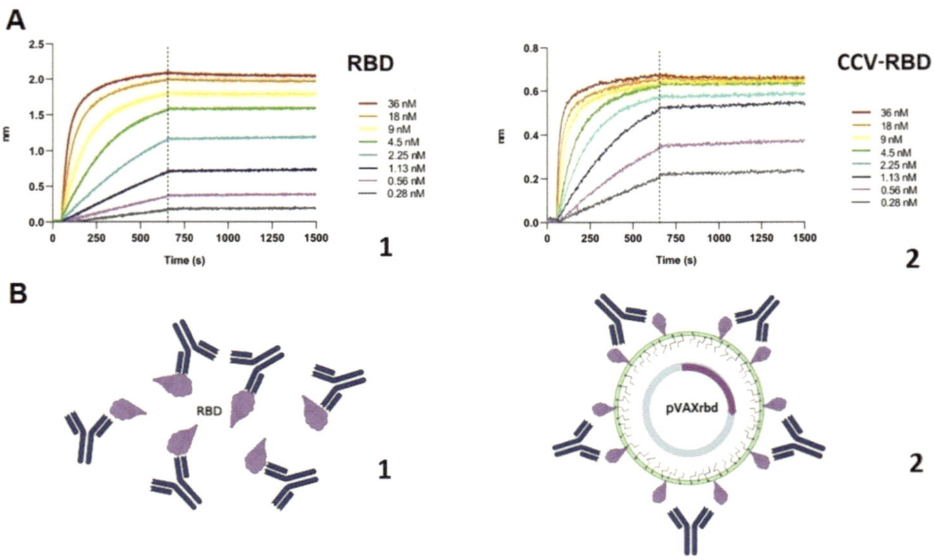

Figure 8.
(A1) biolayer interferometry assay of the RBD to SARS-CoV-2 nAb binding. (A2) biolayer interferometry assay of the CCV-RBD to SARS-CoV-2 nAb binding. (B1, B2) schematic representation of the interaction between RBD protein or CCV-RBD particle and antibody. We express our gratitude to Taranin A.V. and Baranov K.O. For the biolayer interferometry performing.

exposure on the particle surface (**Figure 8**) and its spatial multimerization, which can potentially significantly enhance the RBD-specific humoral immune response.

Female BALB/c mice weighing 16–18 g were used to evaluate the immunogenicity of the construct. All experiments with animals were performed in compliance with the principles of humanity according to the protocols approved by the Bioethics Committee of Vector State Research Center. Mice were divided into groups of eight animals each and immunized as follows: CCV-RBD group - a combined vaccine containing 100 μg of DNA and 50 μg of protein; pVAXrbd group - 100 μg of "naked" DNA vaccine; RBD-PGS group - 50 μg of RBD protein conjugated with polyglucin-spermidine. The intact group contained unimmunized animals. Mice were immunized intramuscularly twice at three-week intervals into the hind thigh. Ten days after the second immunization, blood and spleens were drawn from the animals to examine the virus-specific immune response.

According to the results of the RBD-specific ELISA at the endpoint of the experiment, it can be concluded that the use of the protein as a humoral component significantly increases the average titers of specific antibodies in the animals. Thus, in the group immunized with "naked" pVAXrbd DNA, the titer of specific IgG was 1:2617, in the group that received RBD protein was 1:22280, whereas in the group that received the combined preparation, this figure reached the value 1: 369900 (**Figure 9A**). Thus, the combination of DNA and protein resulted in a synergistic enhancement of the specific humoral immune response compared with the administration of the individual components: 17-fold relative to protein alone, 140-fold relative to "naked" DNA ($p < 0.01$).

Sera from all groups except intact mice and mice immunized with "naked" pVAXrbd demonstrated the ability to neutralize the nCoV/Victoria/1/2020 strain on cell culture *in vitro*. However, in the CCV-RBD group, the average neutralizing titer was 1:376, which was significantly higher than the titers shown in the RBD group (1:21) ($p < 0.05$) (**Figure 9B**).

The synergistic effect with respect to the humoral response may Indicate the effective capture of CCV virus-like particles by APCs, as well as the stimulation of the T-helper response due to the DNA component. Multiple representations of RBD protein on the particle surface may also play a role in enhancing the humoral immune response to CCV-RBD.

Thus, it has been shown that administration of the DNA/protein vaccine results in a stronger humoral response compared to administration of DNA alone or protein alone.

The number of splenocytes producing IFN-γ in response to specific stimulation by the peptide pool of RBD protein was assessed using the ELISpot method. Splenocytes were isolated from the spleens of the animals participating in the experiment. It was shown that a high rate of cellular immunity was recorded in the group immunized with the DNA vaccine in the absence of protein (**Figure 9C**). The addition of the protein component to the vaccine resulted in the formation of a lower cellular response, which may be due to the increased uptake of the combined particles by APC and their conduction through the lysosomal cleavage pathway with a low percentage of escaping DNA and its subsequent penetration into the cell nucleus.

The pVAXrbd DNA vaccine and the combined CCV-RBD vaccine were chosen for protection studies using the SARS-CoV-2 strain hCoV-19/Russia/SA-17620-080521/2021, belonging to the P.1 line (gamma variant). Protein conjugate of RBD and polyglucin-spermidine was used as a protein control. Infection was performed 2 weeks after the second immunization, and the intranasal dose was 50 ID_{50}. As shown in **Figure 9D**, minimal viral load was detected in mice immunized with the combined

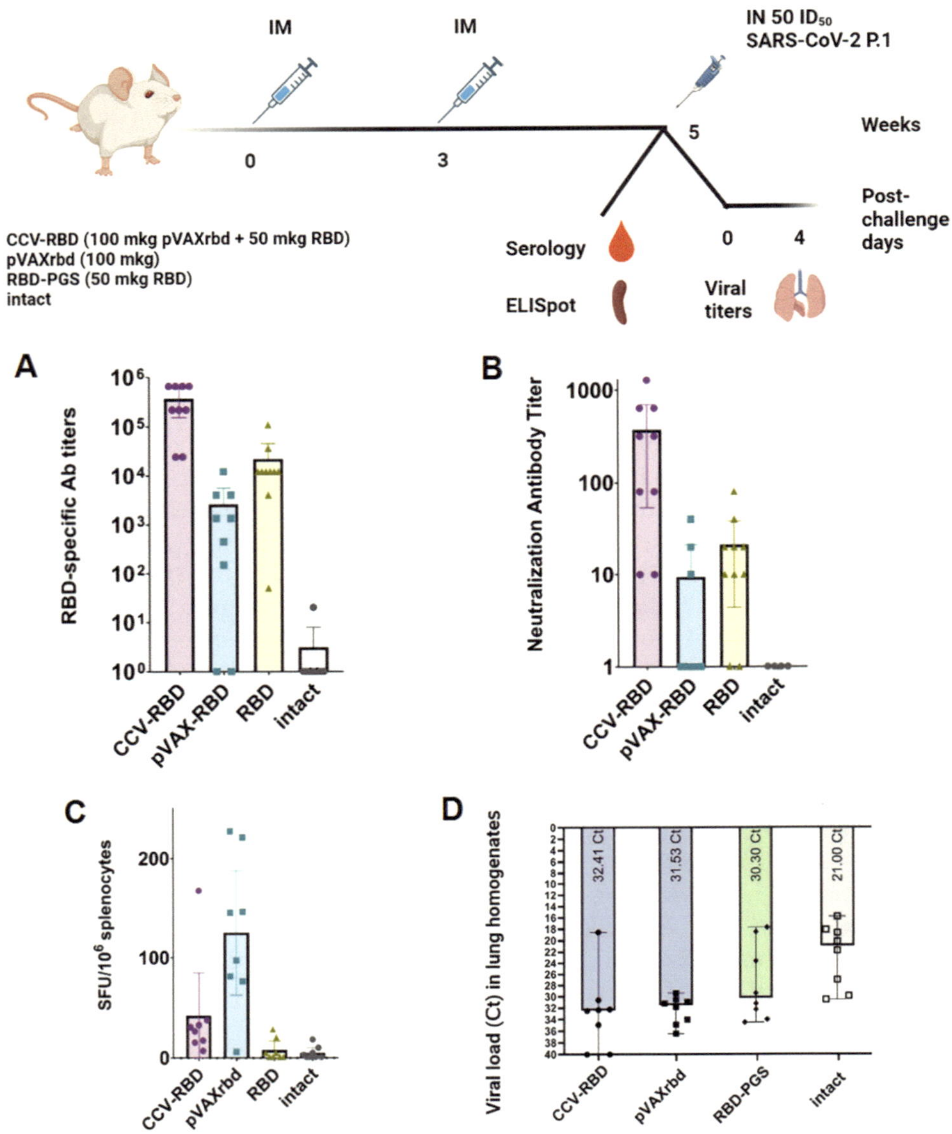

Figure 9.
Design of the experiment and analysis of immunogenicity and protectivity. (A) RBD-specific IgG titers were determined by ELISA. (B) Viral neutralization of murine sera was determined using SARS-CoV-2 nCoV/Victoria/1/2020 strain (100 $TCID_{50}$). (C) T-cell response: Number of splenocytes producing IFN-γ in response to specific stimulation, per 10^6 cells determined using ELISpot. (D) Viral load in lung tissues of BALB/c mice on day 4 after infection with gamma variant SARS-CoV-2.

CCV-RBD vaccine. The decrease in viral RNA amount compared to the control was 3.43 lg (32.41 Ct vs. 21.00 Ct ($p < 0.01$)). Decrease in the amount of viral RNA in lungs of mice immunized with the pVAXrbd DNA vaccine and RBD protein was 3.18 and 2.79 lg, respectively, compared with control (31.53 and 30.30 Ct vs. 21.00 Ct ($p < 0.05$)).

Thus, all experimental groups showed a tendency to reduce the viral load, but the vaccination with the combined DNA/protein vaccine provided good protection to the lung tissues.

4. Conclusion

DNA vaccines are a promising vaccine platform that has been rapidly developing in recent years. Five vaccines are currently approved in the U.S. veterinary medicine [60–64]. DNA vaccines against a number of viral infections, cancers, and autoimmune diseases are in clinical trials nowadays [65].

The rapid worldwide spread of SARS-CoV-2 has led to the unprecedented development of various vaccine creation platforms, such as nucleic acid-based and vector vaccines, and to the approval of the first human DNA vaccine, ZyCoV-D. However, problems including the low immunogenicity of DNA vaccines still need to be solved.

In our work, we proposed a number of strategies to improve the immunogenicity of DNA vaccines against COVID-19: one encoding the receptor-binding domain of protein S of the SARS-CoV-2 and the other encoding an artificial polyepitope T-cell immunogen.

The attachment of the secretory domain to the RBD (in pVAXrbd) ensured a high level of secretion of the protein product from the cell and, consequently, a level of humoral immunity. Addition of ubiquitin to the artificial polyepitope construct (in pBSI-COV-Ub) provided efficient processing of the product inside the cell and its presentation to MHC I.

The polyglucine-spermidine conjugate was used previously to enhance the immunogenicity of the RBD protein [54], and here, we demonstrate the efficacy of using a PGS conjugate to deliver the pVAX-rbd DNA vaccine.

The use of the cationic polymer polyglucin-spermidine for pVAXrbd DNA-vaccine delivery resulted in an increase in its immunogenicity, especially with respect to the humoral response. The PGS envelope protects the DNA from degradation by nucleases, and the particles formed by the interaction of DNA and PGS are more efficiently taken up by antigen-presenting cells. Nanoparticles composed of such polymers show good biocompatibility, water solubility, biodegradability, and absence of toxicity, and they are also easy and inexpensive to produce and stable over a wide temperature range.

Particles obtained using a polyglucin-spermidine-RBD conjugate containing pVAXrbd inside and RBD protein on the surface demonstrated high immunogenicity. Attachment of RBD protein to the polycationic envelope resulted in synergistic enhancement of the humoral immune response compared to its individual components, as well as in the formation of protective immunity in mice, allowing a significant reduction in the viral load in the lungs of animals after virus challenge.

We hope that the approaches described above will be useful to researchers engaged in the design of DNA vaccines.

Acknowledgements

The study was supported by the Ministry of Science and Higher Education of the Russian Federation (agreement # 075-15-2019-1665).

We would like to sincerely thank the entire team of the Bioengineering Department and the team of the Microorganism Collection Department of the SRC VB "Vector" for their well-coordinated work and the most competent performance of the task set.

Author details

Mariya Borisovna Borgoyakova*, Ekaterina Aleksandrovna Volosnikova,
Aleksander Alekseevich Ilyichev and Larisa Ivanovna Karpenko
State Research Center of Virology and Biotechnology Vector,
Koltsovo, Novosibirsk Region, Russian Federation

*Address all correspondence to: borgoyakova_mb@vector.nsc.ru

References

[1] Momin T, Kansagra K, Patel H, Sharma S, Sharma B, Patel J, et al. Safety and immunogenicity of a DNA SARS-CoV-2 vaccine (ZyCoV-D): Results of an open-label, non-randomized phase I part of phase I/II clinical study by intradermal route in healthy subjects in India. EClinicalMedicine. 2021;**38**:101020. DOI: 10.1016/j.eclinm.2021.101020

[2] Aurisicchio L, Brambilla N, Cazzaniga ME, Bonfanti P, Milleri S, Ascierto PA, et al. A first-in-human trial on the safety and immunogenicity of COVID-eVax, a cellular response-skewed DNA vaccine against COVID-19. Molecular Therapy. 2023;**31**(3):788-800. DOI: 10.1016/j.ymthe.2022.12.017

[3] Tebas P, Roberts CC, Muthumani K, Reuschel EL, Kudchodkar SB, Zaidi FI, et al. Safety and immunogenicity of an anti-Zika virus DNA vaccine. The New England Journal of Medicine. 2021;**385**(12):e35. DOI: 10.1056/NEJMoa1708120

[4] Modjarrad K, Roberts CC, Mills KT, Castellano AR, Paolino K, Muthumani K, et al. Safety and immunogenicity of an anti-Middle East respiratory syndrome coronavirus DNA vaccine: A phase 1, open-label, single-arm, dose-escalation trial. The Lancet Infectious Diseases. 2019;**19**(9):1013-1022. DOI: 10.1016/S1473-3099(19)30266-X

[5] Tebas P, Kraynyak KA, Patel A, Maslow JN, Morrow MP, Sylvester AJ, et al. Intradermal SynCon® Ebola GP DNA vaccine is temperature stable and safely demonstrates cellular and humoral immunogenicity advantages in healthy volunteers. The Journal of Infectious Diseases. 2019;**220**(3):400-410. DOI: 10.1093/infdis/jiz132

[6] Kumru OS, Joshi SB, Smith DE, Middaugh CR, Prusik T, Volkin DB. Vaccine instability in the cold chain: Mechanisms, analysis and formulation strategies. Biologicals. 2014;**42**(5):237-259. DOI: 10.1016/j.biologicals.2014.05.007

[7] Maslow JN, Kwon I, Kudchodkar SB, Kane D, Tadesse A, Lee H, et al. DNA vaccines for epidemic preparedness: SARS-CoV-2 and beyond. Vaccines (Basel). 2023;**11**(6):1016. DOI: 10.3390/vaccines11061016

[8] Mallapaty S, India's DNA. COVID vaccine is a world first - more are coming. Nature. 2021;**597**(7875):161-162. DOI: 10.1038/d41586-021-02385-x

[9] Khobragade A, Bhate S, Ramaiah V, Deshpande S, Giri K, Phophle H, et al. ZyCoV-D phase 3 study investigator group. Efficacy, safety, and immunogenicity of the DNA SARS-CoV-2 vaccine (ZyCoV-D): The interim efficacy results of a phase 3, randomised, double-blind, placebo-controlled study in India. Lancet. 2022;**399**(10332):1313-1321. DOI: 10.1016/S0140-6736(22)00151-9

[10] Megati S, Garcia-Hand D, Cappello S, Roopchand V, Masood A, Xu R, et al. Modifying the HIV-1 env gp160 gene to improve pDNA vaccine-elicited cell-mediated immune responses. Vaccine. 2008;**26**(40):5083-5094. DOI: 10.1016/j.vaccine.2008.03.092

[11] Fath S, Bauer AP, Liss M, Spriestersbach A, Maertens B, Hahn P, et al. Multiparameter RNA and codon optimization: A standardized tool to assess and enhance autologous mammalian gene expression. PLoS One. 2011;**6**(3):e17596. DOI: 10.1371/

journal.pone.0017596/10.1371/annotation/039deb02-bbe7-406c-a876-341cc4f3fefa. Erratum in: PLoS One. 2011;6(3)

[12] Garg S, Oran AE, Hon H, Jacob J. The hybrid cytomegalovirus enhancer/chicken beta-actin promoter along with woodchuck hepatitis virus posttranscriptional regulatory element enhances the protective efficacy of DNA vaccines. Journal of Immunology. 2004;**173**(1):550-558. DOI: 10.4049/jimmunol.173.1.550

[13] Vanniasinkam T, Reddy ST, Ertl HC. DNA immunization using a non-viral promoter. Virology. 2006;**344**(2):412-420. DOI: 10.1016/j.virol.2005.08.040

[14] Kozak M. At least six nucleotides preceding the AUG initiator codon enhance translation in mammalian cells. Journal of Molecular Biology. 1987;**196**(4):947-950. DOI: 10.1016/0022-2836(87)90418-9

[15] Gurunathan S, Klinman DM, Seder RA. DNA vaccines: Immunology, application, and optimization*. Annual Review of Immunology. 2000;**18**:927-974. DOI: 10.1146/annurev.immunol.18.1.927

[16] Barrett LW, Fletcher S, Wilton SD. Regulation of eukaryotic gene expression by the untranslated gene regions and other non-coding elements. Cellular and Molecular Life Sciences. 2012;**69**(21):3613-3634. DOI: 10.1007/s00018-012-0990-9

[17] Gavali S, Liu J, Li X, Paolino M. Ubiquitination in T-cell activation and checkpoint inhibition: New avenues for targeted cancer immunotherapy. International Journal of Molecular Sciences. 2021;**22**(19):10800. DOI: 10.3390/ijms221910800

[18] Zhuo M, Song L, Tang Y, Dai S, Chen X, Yu Y, et al. Vaccination with ubiquitin-hepatitis B core antigen-cytoplasmic transduction peptide enhances the hepatitis B virus-specific cytotoxic T-lymphocyte immune response and inhibits hepatitis B virus replication in transgenic mice. Molecular Medicine Reports. 2015;**12**(3):3591-3598. DOI: 10.3892/mmr.2015.3834

[19] Wadhwa A, Aljabbari A, Lokras A, Foged C, Thakur A. Opportunities and challenges in the delivery of mRNA-based vaccines. Pharmaceutics. 2020;**12**(2):102. DOI: 10.3390/pharmaceutics12020102

[20] Kraynyak KA, Blackwood E, Agnes J, Tebas P, Giffear M, Amante D, et al. SARS-CoV-2 DNA vaccine INO-4800 induces durable immune responses capable of being boosted in a phase 1 open-label trial. The Journal of Infectious Diseases. 2022;**225**(11):1923-1932. DOI: 10.1093/infdis/jiac016

[21] Conforti A, Marra E, Palombo F, Roscilli G, Ravà M, Fumagalli V, et al. COVID-eVax, an electroporated DNA vaccine candidate encoding the SARS-CoV-2 RBD, elicits protective responses in animal models. Molecular Therapy. 2022;**30**(1):311-326. DOI: 10.1016/j.ymthe.2021.09.011

[22] Ahn JY, Lee J, Suh YS, Song YG, Choi YJ, Lee KH, et al. Safety and immunogenicity of two recombinant DNA COVID-19 vaccines containing the coding regions of the spike or spike and nucleocapsid proteins: An interim analysis of two open-label, non-randomised, phase 1 trials in healthy adults. Lancet Microbe. 2022;**3**(3):e173-e183. DOI: 10.1016/S2666-5247(21)00358-X

[23] Nakagami H, Hayashi H, Sun J, Yanagida Y, Otera T, Nakagami F, et al.

Phase I study to assess the safety and immunogenicity of an intradermal COVID-19 DNA vaccine administered using a pyro-drive jet injector in healthy adults. Vaccines (Basel). 2022;**10**(9):1427. DOI: 10.3390/vaccines10091427

[24] Kim WJ, Roberts CC, Song JY, Yoon JG, Seong H, Hyun HJ, et al. Safety and immunogenicity of the bi-cistronic GLS-5310 COVID-19 DNA vaccine delivered with the GeneDerm suction device. International Journal of Infectious Diseases. 2023;**128**:112-120. DOI: 10.1016/j.ijid.2022.12.037

[25] Karpenko LI, Danilenko AV, Bazhan SI, Danilenko ED, Sysoeva GM, Kaplina ON, et al. Attenuated salmonella enteritidis E23 as a vehicle for the rectal delivery of DNA vaccine coding for HIV-1 polyepitope CTL immunogen. Microbial Biotechnology. 2012;**5**(2):241-250. DOI: 10.1111/j.1751-7915.2011.00291.x. Epub 2011 Sep 6

[26] Wyszyńska A, Kobierecka P, Bardowski J, Jagusztyn-Krynicka EK. Lactic acid bacteria–20 years exploring their potential as live vectors for mucosal vaccination. Applied Microbiology and Biotechnology. 2015;**99**(7):2967-2977. DOI: 10.1007/s00253-015-6498-0

[27] Zhu D, Mengyue M, Qimuge A, Bilige B, Baiyin T, Temuqile T, et al. Oral delivery of SARS-CoV-2 DNA vaccines using attenuated salmonella typhimurium as a carrier in rat. Molecular Genetics, Microbiology and Virology. 2022;**37**(3):159-166. DOI: 10.3103/S0891416822030107

[28] Yurina V. Live bacterial vectors-a promising DNA vaccine delivery system. Medical Sciences (Basel, Switzerland). 2018;**6**(2):27. DOI: 10.3390/medsci6020027

[29] Li Y, Wang J, Gao Y, Zhu J, Wientjes MG, Au JL. Relationships between liposome properties, cell membrane binding, intracellular processing, and intracellular bioavailability. The AAPS Journal. 2011;**13**(4):585-597. DOI: 10.1208/s12248-011-9298-1

[30] Franck CO, Fanslau L, Bistrovic Popov A, Tyagi P, Fruk L. Biopolymer-based carriers for DNA vaccine design. Angewandte Chemie (International Ed. in English). 2021;**60**(24):13225-13243. DOI: 10.1002/anie.202010282

[31] GeneOptimizer process for successful gene optimization. Available from: https://www.thermofisher.com/ru/en/home/life-science/cloning/gene-synthesis/geneart-gene-synthesis/geneoptimizer.html [Assessed: May 07, 2023]

[32] JAVA codon adaptation tool. Available from: http://www.jcat.de [Assessed: May 07, 2023]

[33] Borgoyakova MB, Karpenko LI, Rudometov AP, Shanshin DV, Isaeva AA, Nesmeyanova VS, et al. Immunogenic properties of the DNA construct encoding the receptor-binding domain of the SARS-CoV-2 spike protein. Molecular Biology. 2021;**55**(6):889-898. DOI: 10.1134/S0026893321050046

[34] Borgoyakova MB, Karpenko LI, Rudometov AP, Volosnikova EA, Merkuleva IA, Starostina EV, et al. Self-assembled particles combining SARS-CoV-2 RBD protein and RBD DNA vaccine induce synergistic enhancement of the humoral response in mice. International Journal of Molecular Sciences. 2022;**23**(4):2188. DOI: 10.3390/ijms23042188

[35] Karpenko LI, Apartsin EK, Dudko SG, Starostina EV, Kaplina ON, Antonets DV, et al. Cationic polymers for the delivery of the Ebola DNA vaccine

encoding artificial T-cell Immunogen. Vaccines (Basel). 2020;**8**(4):718. DOI: 10.3390/vaccines8040718

[36] Karpenko LI, Ilyichev AA, Eroshkin AM, Lebedev LR, Uzhachenko RV, Nekrasova NA, et al. Combined virus-like particle-based polyepitope DNA/protein HIV-1 vaccine design, immunogenicity and toxicity studies. Vaccine. 2007;**25**(21):4312-4323. DOI: 10.1016/j.vaccine.2007.02.058

[37] Kisakov DN, Kisakova LA, Borgoyakova MB, Starostina EV, Taranov OS, Ivleva EK, et al. Optimization of In vivo electroporation conditions and delivery of DNA vaccine encoding SARS-CoV-2 RBD using the determined protocol. Pharmaceutics. 2022;**14**(11):2259. DOI: 10.3390/pharmaceutics14112259

[38] Ju B, Zhang Q, Ge J, Wang R, Sun J, Ge X, et al. Human neutralizing antibodies elicited by SARS-CoV-2 infection. Nature. 2020;**584**(7819):115-119. DOI: 10.1038/s41586-020-2380-z

[39] Rogers TF, Zhao F, Huang D, Beutler N, Burns A, He WT, et al. Isolation of potent SARS-CoV-2 neutralizing antibodies and protection from disease in a small animal model. Science. 2020;**369**(6506):956-963. DOI: 10.1126/science.abc7520

[40] Shi R, Shan C, Duan X, Chen Z, Liu P, Song J, et al. A human neutralizing antibody targets the receptor-binding site of SARS-CoV-2. Nature. 2020;**584**(7819):120-124. DOI: 10.1038/s41586-020-2381-y

[41] Yu J, Tostanoski LH, Peter L, Mercado NB, McMahan K, Mahrokhian SH, et al. DNA vaccine protection against SARS-CoV-2 in rhesus macaques. Science. 2020;**369**(6505):806-811. DOI: 10.1126/science.abc6284

[42] Ravichandran S, Coyle EM, Klenow L, Tang J, Grubbs G, Liu S, et al. Antibody signature induced by SARS-CoV-2 spike protein immunogens in rabbits. Science Translational Medicine. 2020;**12**(550):eabc3539. DOI: 10.1126/scitranslmed.abc3539

[43] Liu L, Wang P, Nair MS, Yu J, Rapp M, Wang Q, et al. Potent neutralizing antibodies against multiple epitopes on SARS-CoV-2 spike. Nature. 2020;**584**(7821):450-456. DOI: 10.1038/s41586-020-2571-7

[44] Quinlan B, Mou H, Zhang L, Guo Y, He W, Ojha A, et al. The SARS-CoV-2 receptor-binding domain elicits a potent neutralizing response without antibody-dependent enhancement. bioRxiv. 2020. DOI: 10.1101/2020.04.10.036418

[45] Ng OW, Chia A, Tan AT, Jadi RS, Leong HN, Bertoletti A, et al. Memory T cell responses targeting the SARS coronavirus persist up to 11 years post-infection. Vaccine. 2016;**34**(17):2008-2014. DOI: 10.1016/j.vaccine.2016.02.063

[46] Chen J, Huang B, Deng Y, Wang W, Zhai C, Han D, et al. Synergistic immunity and protection in mice by co-immunization with DNA vaccines encoding the spike protein and other structural proteins of SARS-CoV-2. Vaccines (Basel). 2023;**11**(2):243. DOI: 10.3390/vaccines11020243

[47] Mahajan S, Kode V, Bhojak K, Karunakaran C, Lee K, Manoharan M, et al. Immunodominant T-cell epitopes from the SARS-CoV-2 spike antigen reveal robust pre-existing T-cell immunity in unexposed individuals. Scientific Reports. 2021;**11**(1):13164. DOI: 10.1038/s41598-021-92521-4

[48] Bhattacharya M, Sharma AR, Ghosh P, Lee SS, Chakraborty C. A next-generation vaccine candidate

using alternative epitopes to protect against Wuhan and all significant mutant variants of SARS-CoV-2: An immunoinformatics approach. Aging and Disease. 2021;**12**(8):2173-2195. DOI: 10.14336/AD.2021.0518

[49] Safavi A, Kefayat A, Mahdevar E, Abiri A, Ghahremani F. Exploring the out of sight antigens of SARS-CoV-2 to design a candidate multi-epitope vaccine by utilizing immunoinformatics approaches. Vaccine. 2020;**38**(48):7612-7628. DOI: 10.1016/j.vaccine.2020.10.016

[50] Wang Q, Lei C, Wan H, Liu Q. Improved cellular immune response elicited by a ubiquitin-fused DNA vaccine against mycobacterium tuberculosis. DNA and Cell Biology. 2012;**31**(4):489-495. DOI: 10.1089/dna.2011.1309

[51] Reguzova A, Antonets D, Karpenko L, Ilyichev A, Maksyutov R, Bazhan S. Design and evaluation of optimized artificial HIV-1 poly-T cell-epitope immunogens. PLoS One. 2015;**10**(3):e0116412. DOI: 10.1371/journal.pone.0116412

[52] Borgoyakova MB, Karpenko LI, Rudometov AP, Starostina EV, Zadorozhny AM, Kisakova LA, et al. Artificial COVID-19 T cell immunogen. Bulletin of Experimental Biology and Medicine. 2023;**175**(2):767-762. DOI: 10.47056/0365-9615-2023-175-6-767-772

[53] Bolhassani A, Javanzad S, Saleh T, Hashemi M, Aghasadeghi MR, Sadat SM. Polymeric nanoparticles: Potent vectors for vaccine delivery targeting cancer and infectious diseases. Human Vaccines & Immunotherapeutics. 2014;**10**(2):321-332. DOI: 10.4161/hv.26796

[54] Volosnikova EA, Merkuleva IA, Esina TI, Shcherbakov DN, Borgoyakova MB, Isaeva AA, et al. SARS-CoV-2 RBD conjugated to polyglucin, spermidine, and dsRNA elicits a strong immune response in mice. Vaccines (Basel). 2023;**11**(4):808. DOI: 10.3390/vaccines11040808

[55] Karpenko LI, Lebedev LR, Ignatyev GM, Agafonov AP, Poryvaeva VA, Pronyaeva TR, et al. Construction of artificial virus-like particles exposing HIV epitopes, and the study of their immunogenic properties. Vaccine. 2003;**21**(5-6):386-392. DOI: 10.1016/s0264-410x(02)00406-1

[56] Karpenko LI, Bazhan SI, Bogryantseva MP, Ryndyuk NN, Ginko ZI, Kuzubov VI, et al. Results of phase I clinical trials of a combined vaccine against HIV-1 based on synthetic polyepitope immunogens. Russian Journal of Bioorganic Chemistry. 2016;**42**:170-182. DOI: 10.1134/S1068162016020060

[57] Singh A. Eliciting B cell immunity against infectious diseases using nanovaccines. Nature Nanotechnology. 2021;**16**(1):16-24. DOI: 10.1038/s41565-020-00790-3

[58] Lebedev LR, Karpenko LI, Poryvaeva VA, Azaev MS, Ryabchikova EI, Gileva IP, et al. Construction of virus-like particles exposing HIV-1 epitopes. Molecular Biology. 2000;**34**:413-417. DOI: 10.1007/BF02759674

[59] Lebedev LR, Goncharova EP, Sizov AA, Bulychev LE, Odegov AM, Ryzhikov AB. Experimental molecular design of combined vaccines. Molecular Biology. 2003;**37**:464-467. DOI: 10.1023/A:1024255814811

[60] Davidson AH, Traub-Dargatz JL, Rodeheaver RM, Ostlund EN, Pedersen DD, Moorhead RG, et al. Immunologic responses to West Nile virus in vaccinated and clinically

affected horses. Journal of the American Veterinary Medical Association. 2005;**226**(2):240-245. DOI: 10.2460/javma.2005.226.240

[61] Garver KA, LaPatra SE, Kurath G. Efficacy of an infectious hematopoietic necrosis (IHN) virus DNA vaccine in Chinook Oncorhynchus tshawytscha and sockeye O. Nerka Salmon. Diseases of Aquatic Organisms. 2005;**64**(1):13-22. DOI: 10.3354/dao064013

[62] Bergman PJ, Camps-Palau MA, McKnight JA, Leibman NF, Craft DM, Leung C, et al. Development of a xenogeneic DNA vaccine program for canine malignant melanoma at the animal medical Center. Vaccine. 2006;**24**(21):4582-4585. DOI: 10.1016/j.vaccine.2005.08.027

[63] Clynav Salmon pancreas disease vaccine. Available from: https://www.ema.europa.eu/en/medicines/veterinary/EPAR/clynav [Assessed: May 07, 2023]

[64] Jazayeri SD, Poh CL. Recent advances in delivery of veterinary DNA vaccines against avian pathogens. Veterinary Research. 2019;**50**(1):78. DOI: 10.1186/s13567-019-0698-z

[65] DNA vaccines in clinical research. Available from: https://clinicaltrials.gov/search?intr=DNA%20Vaccine [Assessed: May 07, 2023]

Section 3

Molecular Approaches and Characteristics

Chapter 4

Useful Molecular and Cytogenetic Approaches in Population Genetics Studies of Pine Species

Ana Carvalho, Maria João Gaspar, Alexandra Dias, José Luís Lousada, Maria Emília Silva and José Lima-Brito

Abstract

In the last decade, we characterised Portuguese populations of *Pinus sylvestris* L., *Pinus nigra* Arnold and *Pinus pinaster* Aiton by using different molecular and cytogenetic approaches. DNA markers helped assess intra- and inter-population genetic variability, extrapolation of phylogenies, provenances and/or infraspecific taxonomy. Quantitative real-time polymerase chain reaction (qRT-PCR), Classical Cytogenetics techniques and/or Comet assay were valuable to assign origins, infraspecific *taxa* or populations more tolerant to water stress. Seed germination tests coupled with the cytogenetic analysis of root cell division provided clues about the impact of fire recurrence on the natural regeneration ability and root growth, respectively. Molecular Cytogenetics contributed to detecting chromosomal anomalies commonly detected in individuals from peripheral populations relative to the species' natural distribution area. The studies briefly described in this chapter integrated multidisciplinary R&D projects whose assembled results allowed inferences about the adaptive potential of the analysed *Pinus* spp. and relevant information for the definition of strategies concerning germplasm conservation, management, use, and, ultimately, genetic improvement.

Keywords: adaptation, cytogenetics, DNA markers, genetic resources, *Pinus* spp.

1. Introduction

Pine species are an essential component of the European forests. The distribution of pine forests depends on climate change, human action, refuges location, geographical barriers and largely, from their ability of natural regeneration and adaptive potential, with the latter being highly influenced by the tree's genetic diversity [1, 2].

In Portugal (Southwest of Europe), the lowland Mediterranean *Pinus pinaster* Aiton (Maritime pine) is the dominant pine species, highly economically important for resin and timber production. The highland Mediterranean *Pinus sylvestris* L. (Scots pine) and *Pinus nigra* Arnold (Black pine) can be found in scattered small-sized populations distributed in the North and Centre of Portugal. Scots pine and Black pine also have economic and ecological importance in Portugal and other European forest ecosystems [3–5].

For years, the presence of natural *P. sylvestris* (autochthonous) stands at 'Serra do Gerês' (NW of Portugal) was uncertain [6–10]. It was believed that only artificial *P. sylvestris* stands resulting from planting towards timber production and prevention of soil erosion existed [8]. Such suspicions arose from the lack of historical background for *P. sylvestris*. The exact origin of the plant material used in the planting of the artificial (allochthonous) stands of *P. sylvestris* and *P. nigra* was also unclear. Nonetheless, the studies our team developed so far clarified these issues partially. Through this chapter, we will present a brief description of the molecular and cytogenetic studies that we developed during the last decade, in the Portuguese populations of *P. sylvestris*, *P. nigra* and *P. pinaster*, with the aid and collaboration of various research colleagues, students and technicians from UTAD and other national and international institutions, in the scope of different multidisciplinary R&D projects.

2. Methodology

2.1 DNA markers

For the molecular characterisation of different Portuguese pine populations, as dominant DNA markers, we used the: (i) inter-simple sequence repeat (ISSR) [11]; (ii) random amplified polymorphic DNA (RAPD) [12, 13]; and (iii) start codon targeted (SCoT) [14] markers. For the amplification of ISSRs and RAPDs, a concentration of 10 – 25 ng/µL of template DNA was used. The amplification of SCoTs was performed with 60 ng/µL of genomic DNA. The initial concentration of the 10-mer oligonucleotides, SSR and SCoT primers for the amplification of RAPDs, ISSRs and SCoTs was 5 µM. Additional details of reaction mixtures can be found in [15, 16]. For the molecular characterisation of the Portuguese pine populations with ISSRs, RAPDs and SCoTs, we used primers previously developed by others, whose references, along with the primer sequences and amplification conditions used in our previous studies, are presented in **Table 1**.

Table 2 presents the sequences of the SSR primers used to amplify nuclear SSRs (nSSRs), expressed sequence tag-SSRs (EST-SSRs) and/or chloroplastidial-SSRs (cpSSRs) by individual or multiplex PCR in *P. sylvestris* [18] and/or *P. nigra* individuals [19].

SSR Primer [Sequence 5′ → 3′] (UBC Primer Set #9, NAPS Unit Standard Primers)	**Amplification conditions in the three *Pinus* spp.**	
817 [$(CA)_8A$]		
827 [$(AC)_8G$]		
834 [$(AG)_8YT$]*	Initial denaturation step: 94°C – 5 min;	
835 [$(AG)_8YC$]*	Denaturation: 94°C – 30 s;	45 cycles
836 [$(AG)_8YA$]*	Primer annealing: 52°C – 45 s;	
841 [$(GA)_8YC$]*	Extension: 72°C – 2 min;	
850 [$(GT)_8YC$]*	Final extension: 72°C – 5 min.	
866 [$(CTC)_6$]		

RAPD 10-mer primer [Sequence 5′ → 3′] (Operon Technologies, Inc. – Qiagen)	Amplification conditions in *P. sylvestris*	
OPA-9 [GGGTAACGCC]		
OPA-10 [GTGATCGCAG]	Initial denaturation step: 94°C – 6 min;	
OPA-15 [TTCCGAACCC]		
OPA-17 [GACCGCTTGT]	Denaturation: 94°C – 1 min; Primer annealing: 36°C – 1 min; Extension: 72°C – 2 min;	40 cycles
OPA-19 [CAAACGTCGG]		
OPH-4 [GGAAGTCGCC]		
OPH-7 [CTGCATCGTG]	Final extension: 72°C – 6 min.	
OPH-19 [CTGACCAGCC]		
SCoT primers [Sequence 5′ → 3′] [14]	**Amplification conditions in *P. nigra* and/or *P. pinaster***	
SCoT 3 [CAACAATGGCTACCACCG]		
SCoT 7 [CAACAATGGCTACCACGG]	Initial denaturation step: 94°C – 3 min;	
SCoT 17 [ACCATGGCTACCACCGAG]		
SCoT 18 [ACCATGGCTACCACCGCC]	Denaturation: 94°C – 1 min; Primer annealing: 50°C – 1 min; Extension: 72°C – 2 min;	35 cycles
SCoT 25 [ACCATGGCTACCACCGGG]		
SCoT 26 [ACCATGGCTACCACCGTC]		
SCoT 27 [ACCATGGCTACCACCGTG]	Final extension: 72°C – 5 min.	
SCoT 31 [CCATGGCTACCACCGCCT]		

Table 1.
Primer sequences used for the amplification of the dominant ISSR, RAPD and SCoT markers in P. sylvestris *[15],* P. nigra *[16] and/or* Pinus pinaster *[17], and respective amplification conditions.*

Details of the fluorescent dyes used to label the forward primer of each pair, as required for the capillary electrophoresis, along with the composition of the reaction mixtures used for individual or multiplex PCR, and the size of the amplified SSR fragments, in both pine species are available in [18, 19].

2.2 Cytogenetic techniques

Root meristematic cells of *P. sylvestris*, *P. nigra* and *P. pinaster* were cytogenetically analysed by using Classical Cytogenetics, conventional fluorescence *in situ* hybridisation (FISH) and non-denaturing FISH (ND-FISH). Upon germination, the collection of roots with 1.5-cm length allows the observation of mitotic dividing cells (**Figure 1**). When the goal of the study is the evaluation of the root mitotic cell cycle, the collected roots are immediately fixed in a solution of absolute ethanol and acetic in the proportion of 3:1 (*v/v*) and then stained with 2% aceto-carmine for 48 h, to enable the observation of interphase and dividing cells into all mitotic phases for further determination of the cytogenetic parameters, mitotic index (MI) and dividing cells with anomalies (DCA), both expressed in percentage [26, 27] (**Figure 1**). On the other hand, for the counting of chromosomes or physical mapping of repetitive DNA sequences such as SSRs or rDNA *loci*, root treatment in ice-cold water (24h – 0°C) or metaphase arresting agents (e.g. 0.05% colchicine or 2 mM 8-hydroxyquinoline) at 4°

nSSR primer [Sequence 5′ → 3′] [20]	**Amplification conditions in *P. sylvestris***	
SPAC11.4 – F [TCACAAAACACGTGATTCACA]	Initial denaturation step: 98°C – 1 min;	
SPAC11.4 – R [GAAAATAGCCCTGTGTGAGACA]	Denaturation: 98°C – 10 s; Primer annealing: 55°C – 10 s; Extension: 72°C – 20 s;	10 cycles (decrease of 1 °C in the primer anneling temperature per cycle)
SPAC11.8 – F [AGGGAGATCAATAGATCATGG]	Denaturation: 98°C – 10 s; Primer annealing: 50°C – 10 s; Extension: 72°C – 20 s;	40 cycles
SPAC11.8 – R [CAGCCAAGACATCAAAAATG]	Final extension: 72°C – 3 min.	
SPAC12.5 – F [CTTCTTCACTAGTTTCCTTTGG]	Initial denaturation step: 98°C – 1 min;	
SPAC12.5 – R [TTGGTTATAGGCATAGATTGC]	Denaturation: 98°C – 10 s; Primer annealing: 60°C – 10 s; Extension: 72°C – 20 s;	10 cycles (decrease of 1 °C in the primer anneling temperature per cycle)
SPAG7.14 – F [TTCGTAGGACTAAAAATGTGTG]	Denaturation: 98°C – 10 s; Primer annealing: 60°C – 10 s; Extension: 72°C – 20 s;	40 cycles
SPAG7.14 – R [CAAAGTGGATTTTGACCG]	Final extension: 72°C – 3 min.	
nSSR primer [Sequence 5′ → 3′] [20–23]	**Amplification conditions in *P. nigra***	
SPAG7.14 – F [TTCGTAGGACTAAAAATGTGTG]		
SPAG7.14 – R [CAAAGTGGATTTTGACCG]		
pn6360 – F [CAACTTCCTTCACCTGGCAC]		
pn6360 – R [GACCCTTCTCAGCATCAACAC]		
pn7754 – F [TCAAGCTAATGCTGGGAACTC]		
pn7754 – R [GTTGGTCGTGGTGACAAAGG]		

pn4379 – F [ACAGGCACACAATTTGGTCC]	
pn4379 – R [CATCATCTGCAGAAGGGAAGAC]	
pn2246 – F [TCATGTCCAGCTTTGGTTGC]	
pn2246 – R [AAGGTTCATTTCCACTCCGG]	Initial denaturation step: 95°C – 5 min;
pn6175 – F [ATTTCCCGCCTACCATTACC]	Denaturation: 95°C – 30 s;
pn6175 – R [GTTACCTGCAATTCGTGTGG]	Primer annealing: 57°C – 90 s; Extension: 72°C – 30 s; (32 cycles)
pn1403 – F [AGGCAATGATCATGTGGGTC]	Final extension: 60°C – 30 min.
pn1403 – R [GGATAGCCTGCAACTCCAATG]	
pn8747 – F [GCATTCTTTCCTTGCTCTCG]	
pn8747 – R [CTCATCTGCTCCCTCAGTCC]	
pn6266 – F [TGCCTTGCTGATTTATTTCCG]	
pn6266 – R [CACTTTCAGCTACGACCGAC]	
PtTx4001 – F [CTATTTGAGTTAAGAAGGGAGTC]	
PtTx4001 – R [CTGTGGGTAGCATCATC]	
PtTx3107 – F [AAACAAGCCCACATCGTCAATC]	
PtTx3107 – R [TCCCCTGGATCTGAGGA]	
EST-SSR primer [Sequence 5′ → 3′] [24]	**Amplification conditions in *P. nigra***
PHA-6062 – F [CTCTCTTTCTCTTCGGTACTCA]	Initial denaturation step: 95°C – 5 min; Denaturation: 95°C – 30 s;
PHA-6062 – R [AAGTTTCAGAGGATTCACACTC]	Primer annealing: 57°C – 90 s; Extension: 72°C – 30 s; (32 cycles) Final extension: 60°C – 30 min.

PHA-4783 – F [AATCTTCTCTTTCTTTGACTGTGT]	
PHA-4783 – R [CATGGGAATGTTAGGAGTGC]	
cpSSR primer [Sequence 5′ → 3′] [25]	**Amplification conditions in *P. sylvestris***
Pt 15169 – F [CTTGGATGGAATAGCAGCC]	Initial denaturation step: 98°C – 1 min; Denaturation: 98°C – 10 s; Primer annealing: 57°C – 10 s; Extension: 72°C – 20 s; Final extension: 72°C – 3 min. (Denaturation, primer annealing and extension: 35 cycles)
Pt 15169 – R [GGAAGGGCATTAAGGTCATTA]	
Pt 26081 – F [CCCGTATCCAGATATACTTCCA]	
Pt 26081 – R [TGGTTTGATTCATTCGTTCAT]	
Pt 30204 – F [TCATAGCGGAAGATCCTCTTT]	
Pt 30204 – R [CGGATTGATCCTAACCATACC]	
Pt 71936 – F [TTCATTGGAAATACACTAGCCC]	
Pt 71936 – R [AAAACCGTACATGAGATTCCC]	
Pt 87268 – F [GCCAGGGAAAATCGTAGG]	
Pt 87268 – R [AGACGATTAGACATCCAACCC]	

Table 2.
Microsatellite (SSR) primers used for nSSR, cpSSR or EST-SSR markers amplification in P. sylvestris *[18] and/or* P. nigra *[19], respective sequences and amplification conditions.*

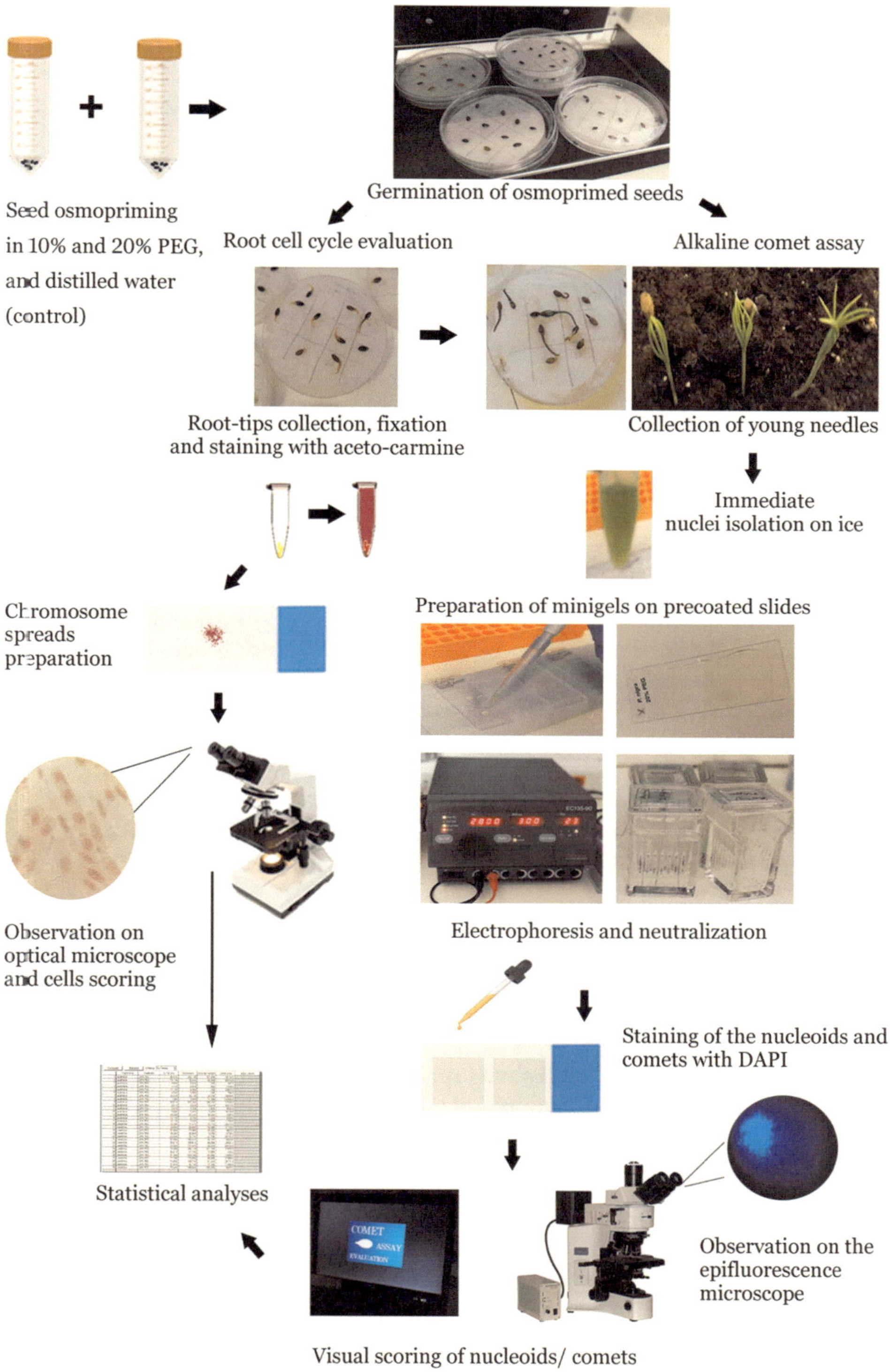

Figure 1.
Schematic representation of the methodology used by [26] to evaluate the effects of induced water stress in the root mitotic cell cycle and nuclear DNA damage in young needles of P. nigra *using Classical Cytogenetics and alkaline Comet assay, respectively.*

C or room temperature for a few hours, before root fixation and/or aceto-carmine staining, should be performed [28]. The treatments of roots with ice-cold water and metaphase-arresting agents inhibit the polymerisation of the microtubules of the mitotic spindle, accumulating C-metaphases that enable the evaluation of individual chromosomes [28].

Fixed root tips of *Pinus* spp. were squashed in the presence of 45% or 60% acetic acid for mitotic spread preparation, and after freezing at −80°C, the glass coverslip was removed, and air-drying was allowed. For the detection of the rDNA *loci* on the *P. sylvestris* chromosomes, conventional FISH was performed with the cloned 45S rDNA probe, pTa71 [29] that despite being isolated in *Triticum aestivum* [30], presents homology with the rDNA *loci* of several plant species. The insert of this clone is an *Eco*-RI fragment of 9 kb comprising one repeat unit of the rRNA genes 18S–5.8S-25S, the intergenic spacer (IGS) and the internal transcribed spacers 1 and 2 (ITS-1 and ITS-2, respectively) that flank the 5.8S rRNA gene [30, 31]. The cloned rDNA sequence pTa71 is usually labelled by nick translation using commercial kits.

For the physical mapping of oligonucleotide SSR sequences on Scots pine chromosomes, the FISH variant technique, ND-FISH, was used [29, 32]. The SSR sequences can be purchased as synthetic oligonucleotides directly labelled in their 5′-end with fluorochromes (e.g. fluorescein or TAMRA) or indirectly labelled with biotin-16-dUTP or digoxigenin-11-dUTP using commercial Random Primed DNA Labelling kits. ND-FISH allows the hybridisation of single-strand DNA oligonucleotide probes (e.g. telomeric, centromeric or SSR sequences) in a few hours since it does not require chromosomal spread pre-treatments [32, 33]. Despite the simplicity of both hybridisation mixture and ND-FISH protocol (**Table 3**), the hybridisation signals present high intensity, specificity and resolution [32, 33]. The non-denaturing conditions preserve the chromosomal morphology [33], enabling successive hybridisations with additional DNA probes by ND-FISH or conventional FISH on the same chromosome spread [34]. Although unnecessary for ND-FISH, chromosome spread pre-treatments (**Table 3**) can be done if successive hybridisations of DNA probes by FISH are envisaged [34].

Unfortunately, the Comet assay is rarely performed in plants when compared to animal species, but the gathered results can complement and enrich Plant Cytogenetics studies as demonstrated before [26] (**Figure 1**). **Figure 1** presents a scheme with the main steps of the alkaline Comet assay that was performed, as far as we know, for the first time, in *P. nigra* [26].

Additional details about the alkaline Comet assay procedure and analysis performed in *P. nigra* can be found in [26].

Main protocol steps	Conventional FISH	ND-FISH
Pre-treatments of the chromosome spreads	• Ageing (dry heat for 2 h) • RNase (1 h) • 2× SSC washes (15 min) • Cytoplasm digestion (Pepsin, 10 min) • 2× SSC washes (15 min) • (Re)fixation of plant material (Paraformaldehyde 10 min) • 2× SSC washes (15 min) • Dehydration in a series of ethanol solutions (70%, 90%, 100%, 15 min) • Air dry	Not required

Main protocol steps	Conventional FISH	ND-FISH
Hybridisation mixture	Formamide	2× SSC
	20× SSC pH 7.0	2 pmol of each oligonucleotide probe
	50% Dextran sulphate	
	Salmon sperm (1 μg/μL)	
	10 mM EDTA pH 8.0	
	10% SDS	
	Probe(s) concentration: 3–7 ng/μL	
	Ultra-pure water	
Chromosomal denaturing	Temperature: ˃ 65°C % Formamide: 50% Final concentration of salt: 2× SSC	Not required
Hybridisation temperature and duration	37°C – overnight	24°C – 2 h
Post-hybridisation washes	• At 42°C; • Washes (5 min each): 2× SSC, twice in 20% formamide/2× SSC; and twice in 0.01 × SSC; • Blocking unspecific sites with Bovine Serum Albumin (BSA) (20 min) where probe molecules bound; • Previous step alternates with washes for 3 × 5 min in detection buffer (4× SSC; Tween 20).	• At room temperature; • Blocking unspecific sites with BSA where probe molecules bound; • Previous step alternates with washes for 3 × 5 min in detection buffer (6× SSC) are performed. .
Detection of hybridisation signal	• Incubation with specific antibodies (37°C, 1 h), unless the probes were directly labelled with fluorochromes. • Previous step alternates with washes for 3 × 5 min in detection buffer (4× SSC; Tween 20).	• Incubation with specific antibodies (37°C, 1 h), unless the probes were directly labelled with fluorochromes. • Previous step alternates with washes for 3 × 5 min in detection buffer (6× SSC).
Chromosomes counterstaining with DAPI. Store at 4°C. Observation on the epifluorescence microscope.		

Table 3.
Main steps of conventional FISH and ND-FISH protocols.

3. (Cyto)molecular studies in *Pinus* spp.

3.1 *Pinus sylvestris*

P. sylvestris (Scots pine) is the conifer with the broadest area of natural distribution, which extends from the British Isles to the Siberian taiga and occupies Southern Europe [35–42]. Portugal constitutes the westernmost limit of the Scots pine natural distribution area.

The various *P. sylvestris* refugees that resulted from the Holocene Epoch greatly influenced the genetic diversity and adaptation of the species found in multiple ecological niches [5, 40]. In the last decades, the increased temperature has enhanced growth and reproduction at the northern limit of its distribution area. In contrast, in its southern boundary, water stress has decreased growth and, in some cases, massive mortality.

In the scope of a multidisciplinary R&D project entitled "*P. sylvestris* in Portugal: the Southwest end or just the end?" (PTDC/AGR-CFL/110988/2009) supported by the FCT – Portuguese Foundation for the Science and Technology and executed at UTAD, a total of nine Scots pine populations were characterised. In the first approach, we performed the molecular characterisation of seven Scots pine populations located in the North and Centre of Portugal, recognised as planted stands and representative of the species distribution in our country. To assess the intra- and inter-population genetic variability, genetic structure and relationships, we used RAPD and ISSR markers [15]. Furthermore, the RAPD and ISSR patterns achieved in the Portuguese samples were compared with others produced in Scots pine individuals from native populations of different countries [15]. The global RAPD and ISSR data analysis revealed higher genetic variability within rather than among populations and a highly differentiated genetic structure [15]. The genetic relationships among the Scots pine populations from different provenances indicated higher similarity between the Portuguese and German individuals, suggesting that the plant material used in the establishment of the *P. sylvestris* stands at the North and Centre of Portugal probably had origin in Central Europe [15]. Scots pine untraceable seeds have been traded and planted across Europe for a long time [43, 44]. The molecular characterisation of Scots populations in different countries can provide valuable information for traceability, certification of forest reproductive material, and definition of germplasm conservation and management strategies [5].

The two putative native Scots pine populations from 'Serra do Gerês' (NW Portugal) [6, 8, 9] were characterised by the codominant, multi-allelic and highly reproducible nSSR and cpSSR markers [18]. The nSSRs are highly efficient for plant genotyping and have been widely used in population genetics studies [5]. The cpSSRs exploit the DNA polymorphism within the chloroplast genome that is haploid and inherited by the male parent in *Pinus* spp.. Since it shows a lower mutation rate than the nuclear genome, it can provide insights about founder effects, genetic drift, historical bottlenecks and other evolutionary events [5, 42, 45]. DNA markers targeting the mitochondrial genome that is maternally inherited and spread by the seeds are also helpful in population genetic studies allowing the exploitation of the history and postglacial recolonisation routes [5 and references therein]. Using nSSR markers, [18] reported higher intra-population polymorphism than among populations. Several studies based on these neutral DNA markers have reported similar results for *Pinus* spp. [5, 19]. The cpSSRs identified distinct haplotypes in the Portuguese populations, resulting in their clustering apart from different European native populations used for comparison [18]. Besides, the two Portuguese populations also showed genetic differentiation between them [18]. Such a study supported the hypothesis of the existence of native Scots pine populations in Portugal and our country as the westernmost limit of the species' natural distribution [18]. The high genetic diversity found in the Portuguese Scots pine populations [15, 18] can explain its adaptive potential to the Mediterranean temperate climate corroborating its broad adaptive potential and phenotypic plasticity, reported mainly for this species with the highest natural distribution area among the conifers [39]. However, as noted earlier for other marginal *P. sylvestris* populations located in the limits of the natural distribution area of the species, such individuals may show phenotypic variations due to the adaptation pressure to a unique environment, along with cytogenetic instability [[46] and references therein]. Therefore, we analysed mitotic chromosome spreads prepared with fixed root-tips upon germination of seeds collected in the native populations of 'Serra do Gerês' [29]. As expected for most of the conifer genera and particularly a species from genus *Pinus* [47], the karyotype of the Scots pine individuals from 'Serra do Gerês' presented 2n = 2x = 24 chromosomes with similar morphology and size (**Figure 2a**).

Previous cytogenetic studies done in Scots pine from marginal populations were based on Classical Cytogenetics techniques and allowed analyses at the morphometric, mitotic, meiotic and nucleolar activity levels [46, 47]. The karyotype of Scots pine was referred to as composed of 11 pairs of metacentric (I – X, XII) and one pair (XI) of submetacentric chromosomes [46, 47]. Mitotic and meiotic instabilities and chromosomal irregularities in individuals from marginal Scots pine populations were reported [46, 47]. To analyse the chromosomes of the Scots pine individuals from 'Serra do Gerês', we used the Molecular Cytogenetics techniques, conventional FISH and ND-FISH, performed with the 45S rDNA pTa71 and 14 SSR sequences, respectively, as probes [29]. The root chromosome spreads showed anomalies such as ring chromosomes, polycentric chromosomes (with additional constrictions beyond the centromere), centric and acentric chromosomal fragments [29] (**Figure 2b** and **c**). These anomalies were similar to those reported earlier for marginal populations of Scots pine and other conifers [[46] and references therein]. The marginal populations are subjected to different selection pressures than those within the central distribution area [29]. The stressful environmental conditions might be responsible for the molecular data achieved in the two native Scots pine populations from 'Serra do Gerês',

Figure 2.
Normal (a) and irregular (b, c) C-metaphase cells of P. sylvestris *(2n = 2× = 24) observed in DAPI-counterstained chromosome spreads, prepared with ice-cold water treated roots, showing: (a) 24 chromosomes with similar size and shape; (b) anomalies like one ring chromosome (arrow) and polycentric chromosomes (arrowheads indicating one of those chromosomes as an example); and (c) centric (arrow) and acentric (arrowhead) chromosomal fragments.*

namely, the high intra-population genetic variability and higher differentiation relative to the populations from Central Europe used for comparison [18]. The high molecular instability can be the origin of the observed cytogenetic irregularities due to the occurrence of DNA strand breakage in response to stress factors.

The pTa71 probe detected 14 rDNA loci on the short and long arms of seven Scots pine chromosome pairs [29]. Among the 14 tested SSR probes, eight of them, based on di-, tri- and tetranucleotide repeat motifs, revealed hybridisation signals on centromeric, pericentromeric, telomeric, subtelomeric and/or interstitial regions of the Scots pine chromosomes [29]. The SSR $(AG)_{10}$ showed a more discriminative hybridisation pattern through the 24 Scots pine chromosomes and allowed their karyotyping [29].

The *Pinus* spp. genomes are physically large, as expressed by their DNA contents [48], and highly complex. The high amount of repetitive DNA sequences of the pine genomes provides a vast adaptive potential, and its analysis can be valuable for evolutionary studies. Hizume et al. [49] demonstrated that using telomeric, centromeric and rDNA sequences as FISH probes constituted reliable cytogenetic markers and allowed the identification of individual chromosomes in four pine species. The widely dispersed and abundant SSRs and retrotransposon-based sequences could also comprise excellent choices for FISH probes enabling additional karyotypes and/or new evolutionary clues among the genomes of pines and other conifers.

Considering the actual climate change scenario and projections, among other extreme environmental episodes, an increase in the frequency and severity of heat waves and drought episodes is expected [50]. A temperature window, a typical pattern of the Mediterranean *Pinus* spp., greatly influences pine seed germination and seedling emergence [51]. On the other hand, water stress, particularly the summer drought, is a limiting factor for seedling survival, reducing the forestry area, changing distribution patterns, compromising the success of afforestation programmes, and, ultimately, the persistence of pine stands [26, 52–54]. When we induced three water availability regimes (watering, moderate water stress and drought) to Scots pine populations from five European provenances ('Gerês', 'Puebla de Lillo', 'Montes Universales', Germany and Sweden), we verified that the Portuguese one, represented by 'Ribeira das Negras' ('Gerês'), had higher tolerance to moderate water stress and drought than the remaining ones [53]. The Portuguese individuals presented the highest stability in photosynthetic reactions and better photochemical and metabolic behaviours under drought [53]. Besides, the relative expression ratio of three water stress-responsive genes (*abaH*, *lp3*, *PAL1*) in individuals from 'Gerês' were lower but increased gradually with the scarcity of water, followed by those of German [53]. These molecular results were exciting regarding our earlier study that assigned Germany as a potential origin of the plant material used to establish the allochthonous Scots pine populations in the North and Centre of Portugal [15]. The Portuguese germplasm of Scots pine, including the native populations of 'Gerês' and the planted ones, have the ability and high adaptive potential to respond to water deficit conditions [53]. On the other hand, the authors verified that the Swedish Scots pine individuals responded faster to moderate and severe water stress revealing the lowest tolerance among the studied provenances [53]. Also, the young needles of the Portuguese Scots pine individuals have physiological and morpho-anatomic parameters more suited to environments with low water availability than those of the remaining provenances [53]. Overall, the genetic, physiological and morpho-anatomic traits evidenced by the 'Gerês' population allowed us to suggest that the Portuguese germplasm is well adapted to the temperate climate and responds better and gradually to water stress relative to the other tested European provenances

[53]. This information is precious concerning the conservation and use of germplasm in abroad regions where the afforestation, adaptation and survival of Scots pine have been or will be compromised by climate change. Likewise, in Portugal, Scots pine can be used for the afforestation of high-altitude areas to which *Pinus pinaster* is not adapted and where the risk of pinewood nematode attack is minor.

Abiotic and biotic stresses and other natural and artificial factors have been declining the forest area throughout Europe, bringing negative consequences to the forestry industry. Concerning the excellent adaptation of Scots pine to the SW end of Europe and the need for alternative resinous species for the wood and resin industry, we characterised the physical, chemical and mechanical properties of wood sampled in individuals from the allochthonous Portuguese *P. sylvestris* populations using X-ray microdensitometry, near-infrared (NIR) spectrometry and bending tests, to infer about its quality [55, 56]. The wood density and radial growth expressed by the ring width of the Portuguese-planted Scots pine individuals were higher than those reported for populations from different northern European regions [55, 56]. Nevertheless, these wood traits did not supplant those exhibited by *P. pinaster*, constituting the main resinous species exploited for industrial purposes in Portugal [55]. The mechanical and physical traits were positively correlated, evidencing that trees with high radial growth can also have excellent mechanical performance [56]. Overall, the Portuguese Scots pine wood evidenced high quality, almost similar to the one of *P. pinaster* [55, 56], reinforcing the need for conservation and use of this germplasm with benefits at the ecological and economic level in Portugal and abroad.

3.2 *Pinus nigra*

The distribution of *P. nigra* (Black pine) in Portugal is restricted to six sites in the North and Centre of the country where adult, regular, mature and planted (allochthonous) stands can be found [16, 19, 57, 58].

Although widely dispersed throughout Europe, the wood quality and growth of *P. nigra* were not very studied. Still, it has the potential to be explored, as revealed earlier by its physical, chemical and mechanical traits, with the advantage that it could be used for the reforestation of mountainous areas where other resinous species with higher economic importance, such as *P. pinaster* are not adapted [57, 58]. In fact, during the 19th century, *P. nigra* was used to reforesting degraded European landscapes due to its adaption to shallow rocky soils and xerophytic sites [59].

The scattered distribution of this pine species through Southern Europe, Asia Minor, Corsica and Sicily islands, and NW Africa led to high inter-population variation at the genetic and biochemical levels and high adaption to different ecological niches [60–63]. Consequently, its infraspecific taxonomic classification has been revised, and the most accepted comprises the existence of six subspecies and some of them with regional varieties. Five *P. nigra* subspecies were assigned to populations distributed through Eurasia: (i) dalmatica (Balkans); (ii) laricio (Corsica, Sicily and Calabria); (iii) nigra (Alps); (iv) pallasiana (Turkey and Crimea) and (v) salzmannii (France and Spain) [10, 62, 64–68], whereas the (vi) subspecies mauretanica (Maire and Peyerimh) Heywood was ascribed to the NW Africa [69].

The origin and infraspecific taxonomy of the plant material used in establishing the Portuguese-planted stand 50–90 years ago are unknown [16]. In the 1980s, a phenological characterisation of the Portuguese *P. nigra* stands classified them as belonging to the subspecies laricio, salzmannii and nigra [70]. Therefore, while the molecular characterisation of the six Portuguese *P. nigra* stands, currently representative of the species distribution in our country, we also analysed samples from different European provenances and with certified infraspecific taxonomy taken into consideration the three subspecies assigned earlier [16, 70]. Firstly, we used the dominant ISSR and Start Codon Targeted (SCoT) markers to evaluate the genetic diversity, relationships and structure and extrapolate the Portuguese populations' infraspecific taxonomy [16]. Both ISSR and SCoT markers revealed higher genetic similarity among the Portuguese populations and those from abroad with provenances belonging to the subspecies laricio [16], one of the subspecies assigned to Portugal by [70]. Nevertheless, when we individually considered the ISSR and SCoT data, the former marker system revealed higher genetic similarity with the variety corsicana, and the second, with the variety calabrica, both from subspecies laricio [16]. The genetic relationships and differentiation of the six Portuguese *P. nigra* populations revealed two main clusters, supporting the possible use of plant material from two different origins in establishing the stands performed in the past century [16]. These results were later reinforced by the use of the codominant nSSRs and cpSSRs, which structured the six Portuguese *P. nigra* populations into two main genetic clusters but with low differentiation ($F_{ST} = 0.04$) [19] matching our previous hypothesis of plant material from two varieties belonging to the same subspecies. Additionally, and similarly to the results achieved with the dominant DNA markers, the SSRs also evidenced a higher genetic diversity within (95%) rather than among (5%) populations [19]. The molecular data achieved in the Portuguese *P. nigra* populations suggested a broad adaptive potential that could be explored if we regard its ability to be used for the reforestation of mountainous areas, reinforcing its ecological value in Portugal and abroad. Also, this species has potential for the forestry industry and can constitute an alternative resinous species considering the results achieved by characterising their wood's physical, chemical and mechanical traits [57, 58]. Some authors pointed out that the natural regeneration in *P. nigra* is complex [59, 71], and it can be even more difficult in dry soils, which considering the climate projections, could be problematic for the species' survival. Hence, we evaluated the water stress tolerance of different infraspecific *taxa* of *P. nigra* based on the analysis of the root mitotic cell cycle and alkaline Comet assay [26] (**Figure 1**). In this latter study, we imbibed *P. nigra* seeds of different provenances with certified infraspecific taxonomy to aqueous solutions of 10% and 20% polyethylene glycol (PEG) (seed osmopriming) during four days to mimic water stress and then allowed their germination on distilled water (**Figure 1**). Roots were collected and immediately fixed upon germination to analyse the effect of induced water stress in the mitotic cell division [26] (**Figure 1**). We verified that this abiotic stress delayed the seed germination, induced mitotic anomalies in the meristematic root cells, and caused nuclear DNA damage as detected in needles of young seedlings weeks after the interruption of the osmotic stress induction [26]. This result suggested a stress memory transmission between plant tissues and can also explain the high rate of seedling mortality widely reported to different *Pinus* spp. [59]. These adverse effects were more pronounced in the subspecies laricio variety corsicana, to which the 20% PEG treatment was lethal [26], which showed high genetic similarity with the Portuguese populations [16]. The evidence of differential water stress tolerance among the studied *P. nigra* infraspecific subspecies and varieties is fascinating. It should be exploited in-depth in additional *taxa,* particularly under climate change scenarios.

The current European *P. nigra* forests are ageing, present decline or constitute fire-prone ecosystems. In the absence of seed trees, the persistence of Black pine stands depends on direct replacement by planting, which is usually chosen but costly, or natural regeneration that, despite preferable, is hard due to various interacting factors

[59]. Natural forestry regeneration is a slow and difficult-to-predict process [72], but it is determinant in the stand persistence [73]. Despite the development of predicting models based on different climate scenarios to simulate regeneration and to help in decision-making concerning forest management [72], the characterisation of marginal pine populations based on the approaches reviewed here is still essential and opportune. The conservation and use of *P. nigra* germplasm with high water stress tolerance, high genetic diversity, known provenance and wide adaptive potential constitute valuable resources to be used in the planting or sowing of Black pine stands in Portugal and abroad.

3.3 *P. pinaster*

Different authors have been reporting the solid genetic control of the resin chemical components and yield in various pine species, including *P. pinaster,* which constitutes the main conifer species exploited for resining in Europe [74, 75]. Efforts have been made to correlate the genotypic and phenotypic data concerning resin yield and quality in different pine species [[75, 76] and references therein]. However, in a first approach, the genetic characterisation of maritime pine stands installed for resining by using molecular markers can be helpful in the selection of elite trees for further establishment of high-resin yielding seed orchards to implement tree breeding programmes towards the improvement of resin yield and quality [17, 74]. A few years ago, our team sampled vascular cambium in 200 adult *P. pinaster* individuals from two populations in the North of Portugal ('Paredes de Coura' and 'Vila Pouca de Aguiar') used for resin production and analysed their genomic DNA with the neutral molecular markers ISSR and SCoT [17]. Both marker systems revealed high intra-population genetic diversity and high inter-population genetic differentiation. The intra-population ISSR and SCoT polymorphism was more elevated in 'Vila Pouca de Aguiar' [17]. Upon comparison of the molecular data with the resin yield values achieved for two consecutive years, we verified that the population of 'Paredes de Coura' that showed lower intra-population ISSR and SCoT polymorphism presented higher resin productivity [17]. Besides, the trees with the highest resin yields gave similar ISSR and SCoT molecular patterns, showing the closest genetic relationships and enabling their selection as elite trees based on these approaches [17]. While low intra-population genetic variation for resin production seems advantageous, high genetic diversity is required to ensure adaptation to stressful factors and species survival. Maritime pine stands are mainly found in the North of Portugal and constitute fire-prone habitats with complex management [77]. This pine species developed evolutionary strategies to cope with fire, such as serotonin protecting the cones' seeds from high temperatures, ensuring their natural regeneration and maintenance of high intra-population genetic diversity [78, 79].To test this latter hypothesis, [60] characterised juvenile *P. pinaster* individuals (5–6 years) naturally regenerated in stands that burned once and twice and compared with adult pine trees from an unburned stand. The authors verified that, in the short term, the Nei's gene diversity (expected heterozygosis) was higher in the individuals from the twice-burned stand and that fire recurrence did not cause genetic erosion. Considering the future increase in wildfires frequency and water deficit, particularly in the Mediterranean region [50], and the influence of these factors on the natural regeneration of *P. pinaster* [80], recently, we analysed seeds from postfire naturally regenerated stands that burned once, twice and three times between the 1990s and 2005 [27]. In the long term, fire recurrence did not influence the seed germination index but delayed the mean germination time significantly, even during recovery [27]. To the same seeds, 10% and

20% polyethylene glycol (PEG) solutions were added before and during germination to analyse how fire recurrence, induced water stress and their interaction impacted seed germination and root cell division [27]. The water stress induction did not significantly influence the germination index of the *P. pinaster* seeds [27]. The fire recurrence negatively impacted the regularity of root mitosis and increased the frequency of irregular cells, mainly in seeds collected in stands that burned twice and thrice [27]. The water stress induced with 10% PEG (osmotic potential: −0.4 MPa) significantly increased the frequency of mitotic anomalies in the root-tip cells [27]. Regardless of the fire regime, seeds exposed to 20% PEG (−0.8 MPa) did not germinate or presented root growth inhibition a few days after the beginning of germination [27]. Anomalies in root cell division might hinder the growth of the radicular system, constituting a significant problem in dry soils.

4. Conclusions

This chapter reviewed molecular and cytogenetic approaches that, individually or combined, integrated multidisciplinary projects and/or postgraduate studies developed at UTAD and focused on Portuguese populations of *P. sylvestris*, *P. pinaster* and *P. nigra*. We are confident that the information we acquired through the last decade concerning genetic diversity, structure and relationships, extrapolation of provenances and infra-specific taxonomy, along with the estimation of their ability to respond to water stress or fire recurrence, can give new insights about the adaptive potential of the Portuguese pine germplasm that can be explored at the national and international levels. The gathered information is helpful for decision-making regarding the definition of strategies for silvicultural practices, conservation and use of genetic resources, and implementation of breeding programmes, particularly for marginal pine populations that inhabit ecosystems in the limits of their natural distribution, experiencing unfavourable climatic conditions that might increase the risk of seed recruitment and competition for resources, threaten the stands and species persistence.

Acknowledgements

The authors acknowledge the contribution of all researchers' colleagues, field technicians and students from UTAD and other national and international institutions involved in the execution of the cited works. Additionally, the authors acknowledge the funding sources that enabled the execution of the cited works, namely: **(i)** the Portuguese Foundation for the Science and Technology (FCT – "Fundação para a Ciência e a Tecnologia") which supported the R&D project PTDC/AGR-CFL/110988/2009 co-financed by the European Fund of Regional Development (FEDER) – COMPETE-QREN program (leaded by J. Lima-Brito); projects UID/AGR/04033/2019, UIDB/04033/2020 and POCI-01-0145-FEDER-006958-FEDER/COMPETE/POCI - Operational Competitiveness and Internationalisation Program attributed to CITAB/UTAD; the postdoctoral grant SFRH/BPD/68932/2010 (attributed to A. Carvalho) and the doctoral grants SFRH/BD/91781/2012 (attributed to A. Dias) and SFRH/BD/98777/2013 (supervised by J. Lima-Brito, J. Lousada and M.J. Gaspar) co-financed by the Social European Fund (FSE) under the POPH-QREN program; **(ii)** the COST Action FP1202 "Strengthening conservation: a key issue for adaptation of marginal/peripheral populations of forest trees to climate change in Europe

(MaP-FGR)" which financed a Short Term Scientific Mission to A. Dias; and **(iii)** Project RESINPROVE supported by EU funds under the Program of Rural Development in the Continent (PRODER – "Programa de Desenvolvimento Rural do Continente"). Ana Carvalho acknowledges the FCT and UTAD for her contract as a researcher under the scope of the D.L. no. 57/2016 and Law no. 57/2017.

The article processing charges of this chapter were supported by national funds provided by the Portuguese Foundation for the Science and Technology (FCT – "Fundação para a Ciência e a Tecnologia") to the research unit CITAB under the project UIDB/04033/2020.

Author details

Ana Carvalho[1*], Maria João Gaspar[1,2], Alexandra Dias[3], José Luís Lousada[4], Maria Emília Silva[4] and José Lima-Brito[1]

1 Department of Genetics and Biotechnology, Plant Cytogenomics Lab, CITAB – Centre for the Research and Technology of Agro-Environmental and Biological Sciences, Inov4Agro – Institute for Innovation, Capacity Building and Sustainability of Agri-Food Production, University of Trás-os-Montes and Alto Douro (UTAD), Vila Real, Portugal

2 CEF – Centre of Forestry Studies, ISA – Higher Institute of Agronomy, University of Lisbon, Lisboa, Portugal

3 CITAB – Centre for the Research and Technology of Agro-Environmental and Biological Sciences, University of Trás-os-Montes and Alto Douro (UTAD), Vila Real, Portugal

4 Department of Forest Sciences and Landscape Architecture, CITAB – Centre for the Research and Technology of Agro-Environmental and Biological Sciences, Inov4Agro – Institute for Innovation, Capacity Building and Sustainability of Agri-Food Production, University of Trás-os-Montes and Alto Douro (UTAD), Vila Real, Portugal

*Address all correspondence to: anacar@utad.pt

References

[1] Figueiral I, Carcaillet C. A review of Late Pleistocene and Holocene biogeography of highland Mediterranean pines (*Pinus* type *sylvestris*) in Portugal, based on wood charcoal. Quaternary Science Reviews. 2005;**24**:2466-2476. DOI: 10.1016/j.quascirev.2005.02.004

[2] Fady B, Esposito E, Abulaila K, et al. Forest genetics research in the Mediterranean basin: Bibliometric analysis, knowledge gaps, and perspectives. Current Forestry Reports. 2022;**8**:277-298. DOI: 10.1007/s40725-022-00169-8

[3] Portoghesi L, Consalvo M, Angelini A, et al. Multifunctional management of mountain reforestations: Thoughts and perspectives from a case study in Central Italy. L'Italia Forestale e Montana/Italian Journal of Forest and Mountain Environments. 2013;**68**: 305-315. DOI: 10.4129/ifm.2013.6.03

[4] Bonavita S, Vendramin GG, Bernardini V, et al. The first SSR-based assessment of genetic variation and structure among *Pinus laricio* Poiret populations within their native area. Plant Biosystems – An International Journal Dealing with All Aspects of Plant Biology. 2016;**2016**(150):1271-1281. DOI: 10.1080/11263504.2015.1027316

[5] Kavaliauskas D, Danusevičius D, Baliuckas V. New insight into genetic structure and diversity of Scots pine (*Pinus sylvestris* L.) populations in Lithuania based on nuclear, chloroplast and mitochondrial DNA markers. Forests. 2022;**13**:1179. DOI: doi.org/10.3390/f13081179

[6] Rivas-Goday S. Apreciation sintetica de los grados de vegetacion de la Serra de Gerês. Agronomia Lusitana. 1959;**12**: 449-480

[7] Tutin TG, Heywood VH, Burges NA, et al. Flora Europaea. Cambridge: Cambridge University Press; 1964

[8] Franco JA, Afonso MLR. Distribuição de Pteridófitos e Gimnospérmicas em Portugal. Reservas e Património Paisagístico, Lisboa: Serviço Nacional de Parques; 1982

[9] Ribeiro O, Lautensach H, Daveau S. Geografia de Portugal. JS da Costa, Lisboa: II. O ritmo Climático e a Paisagem; 1988

[10] Barbéro M, Loisel R, Quézel P, et al. Pines of the Mediterranean Basin. In: Richardson DM, editor. Ecology and Biogeography of *Pinus*. Cambridge: Cambridge University Press; 1998. pp. 153-170

[11] Zietkiewicz E, Rajalski A, Labuda D. Genome fingerprinting by simple sequence repeat (SSR) – anchored polymerase chain reaction amplification. Genomics. 1994;**20**:176-183

[12] Williams JGK, Kubelik AR, Livak KJ, et al. DNA polymorphisms amplified by arbitrary primers are useful as genetic markers. Nucleic Acids Research. 1990; **18**:6531-6535

[13] Welsh J, McClelland M. Fingerprinting genomes using PCR with arbitrary primers. Nucleic Acids Research. 1990;**18**:7213-7218

[14] Collard BCY, Mackill DJ. Start Codon Targeted (SCoT) polymorphism: A simple, novel DNA marker technique for generating gene-targeted markers in plants. Plant Molecular Biology Reporter. 2009;**27**:86-93. DOI: 10.1007/s11105-008-0060-5

[15] Cipriano J, Carvalho A, Fernandes C, et al. Evaluation of genetic diversity of

Portuguese *Pinus sylvestris* L. populations based on molecular data and inferences about the future use of this germplasm. Journal of Genetics. 2013;**92**(2):e41-e48. DOI: 10.1007/s12041-013-0241-3

[16] Dias A, Lemos M, Pavia I, et al. Genetic characterization of Portuguese allochthonous populations of *Pinus nigra* using ISSRs and SCoTs and extrapolation of their infraspecific taxonomy. Physiology and Molecular Biology of Plants. 2019;**25**:799-805. DOI: 10.1007/s12298-019-00649-5

[17] Silva ME, Gaspar MJ, Pires J, et al. RESINPROVE - Contribuição para a melhoria da eficiência, racionalização e expansão da atividade de resinagem. 2013. Projeto no âmbito da medida 4.1 "Cooperação para a inovação", integrada no subprograma 4 "Promoção do conhecimento e desenvolvimento de competências" do Programa de Desenvolvimento Rural do Continente (PRODER). Editor: Vila Real, Portugal: UTAD – Universidade de Trás-os-Montes e Alto Douro. 2018. p. 40

[18] Pavia I, Mengl M, Gaspar MJ, et al. Preliminary evidences of two potential native populations of *Pinus sylvestris* L. in Portugal based on nuclear and chloroplast SSR markers. Austrian Journal of Forest Science. 2014;**131**:S1-S22. Available from: http://www.forestscience.at/fileadmin/user_upload/forestscience/2014/CB1401_Artikel-1.pdf

[19] Dias A, Giovanelli G, Fady B, et al. Portuguese *Pinus nigra* populations: Genetic diversity, structure and relationships inferred by SSR markers. Annals of Forest Science. 2020a;**77**:64. DOI: 10.1007/s13595-020-00967-9

[20] Soranzo N, Provan J. PowellW: Characterization of microsatellite loci in *Pinus sylvestris* L. Molecular Ecology. 1998;**7**:1247-1263

[21] Giovannellii G, Roig A, Spanu I, Vendramin GG. Fady B: A new set of nuclear microsatellites for an ecologically and economically important conifer: The European Black Pine (*Pinus nigra* Arn.). Plant Molecular Biology Reporter. 2017;**35**:379-388. DOI: 10.1007/s11105-017-1029-z

[22] Zhou Y, Bui T, Auckland L, Williams CG. Undermethylated DNA as a source of microsatellites from a conifer genome. Genome. 2002;**45**:91-99. DOI: 10.1139/g01-119

[23] Elsik CG, Williams CG. Low-copy microsatellite recovery from a conifer genome. Theoretical and Applied Genetics. 2001;**103**(8):1189-1195

[24] Leonarduzzi C, Spanu I, Labriola M, et al. Development and characterization of three highly informative EST-SSR multiplexes for Pinus halepensis Mill. and their transferability to other Mediterranean Pines. Plant Molecular Biology Reporter. 2016;**34**:993-1002. DOI: 10.1007/s11105-016-0980-4

[25] Vendramin GG, Lelli L, Ross P, Morgante M. A set of primers for the amplification of 20 chloroplast microsatellites in Pinaceae. Molecular Ecology. 1996;**5**:595-598

[26] Carvalho A, Gaivão I, Lima-Brito J. Seed osmopriming with PEG solutions in seeds of three infraspecific taxa of *Pinus nigra*: Impacts on germination, mitosis and nuclear DNA. Forest Ecology and Management. 2020;**456**:117739. DOI: 10.1016/j.foreco.2019.117739

[27] Ribeiro S, Gaspar MJ, Lima-Brito J, et al. Impact of fire recurrence and induced water stress on seed germination and root mitotic cell cycle of *Pinus pinaster* Aiton. Forests. 2023;**14**:78. DOI: 10.3390/f14010078

[28] Schwarzacher T, Heslop-Harrison JS. Practical *in situ* Hybridization. Oxford, UK: BIOS Scientific Publishers Limited; 2000

[29] Pavia I, Carvalho A, Rocha L, et al. Physical location of SSR regions and cytogenetic instabilities in *Pinus sylvestris* chromosomes revealed by ND-FISH. Journal of Genetics. 2014b;**93**(2): 567-571. DOI: 10.1007/s12041-014-0412-x

[30] Gerlach WL, Bedbrook JR. Cloning and characterization of ribosomal RNA genes from wheat and barley. Nucleic Acids Research. 1979;**7**:1869-1885

[31] Barker RF, Harberd NP, Jarvist MG, Flavell RB. Structure and evolution of the intergenic region in a ribosomal DNA repeat unit of wheat. Journal of Molecular Biology. 1988;**201**:1-17

[32] Cuadrado A, Jouve N. Chromosomal detection of simple sequence repeats (SSRs) using nondenaturing FISH (ND-FISH). Chromosoma. 2010;**119**:495-503

[33] Cuadrado A, Golczyk H, Jouve N. A novel, simple and rapid nondenaturing FISH (ND-FISH) technique for the detection of plants telomeres - potential used and possible target structures detected. Chromosome Research. 2009; **17**:755-762

[34] Delgado A, Carvalho A, Martín AC, Martín A, Lima-brito J. Genomic reshuffling in advanced lines of hexaploid tritordeum. Genetic Resources and Crop Evolution. 2017;**64**:1331-1353. DOI: 10.1007/s10722-016-0439-3

[35] Labra M, Grassi F, Sgorbati S, et al. Distribution of genetic variability in southern populations of Scots pine (*Pinus sylvestris* L.) from the Alps to the Apennines. Flora. 2006;**201**:468-476

[36] Pyhäjärvi T, Salmela MJ, Savolainen O. Colonization routes of *Pinus sylvestris* inferred from distribution of mitochondrial DNA variation. Tree Genetics and Genomes. 2008;**4**:247-254. DOI: 10.1007/s11295-007-0105-1

[37] Pyhäjärvi T. Kujala ST, Savolainen O: 275 years of forestry meets genomics in *Pinus sylvestris*. Evolutionary Applications. 2020;**13**:11-30. DOI: 10.1111/eva.12809

[38] Scalfi M, Piotti A, Rossi M, et al. Genetic variability of Italian southern Scots pine (*Pinus sylvestris* L.) populations: The rear edge of the range. European Journal of Forest Research. 2009;**128**:377-386

[39] Savolainen O, Kujala ST, Sokol C, et al. Adaptive potential of northernmost tree populations to climate change, with emphasis on Scots pine (*Pinus sylvestris* L.). Journal of Heredity. 2011;**2011**(102): 526-536

[40] Matías L, Jump AS. Interactions between growth, demography and biotic interactions in determining species range limits in a warming world: The case of *Pinus sylvestris*. Forest Ecology and Management. 2012;**282**:10-22

[41] Sannikov SN, Sannikova NS, Petrova IV, et al. The hypothesis about the Lofoten Pleistocene Refugium for *Pinus sylvestris* L. Russian Journal of Ecology. 2019;**50**:218-226. DOI: 10.1134/S1067413619030123

[42] Sheller M, Tóth EG, Ciocîrlan E, et al. Genetic diversity and population structure of Scots pine (*Pinus sylvestris* L.) in Middle Siberia. Forests. 2023;**14**: 119. DOI: 10.3390/f14010119

[43] Mátyás C, Ackzell L, Samuel CJA. EUFORGEN Technical Guidelines for Genetic Conservation and Use for Scots

Pine (*Pinus sylvestris*). Rome, Italy: International Plant Genetic Resources Institute; 2004

[44] Myking T, Rusanen M, Steffenrem A, et al. Historic transfer of forest reproductive material in the Nordic region: Drivers, scale and implications. Forestry. 2016;**89**:325-337

[45] Ribeiro MM, Plomion C, Petit RJ, et al. Variation of chloroplast single-sequence repeats in Portuguese maritime pine (*Pinus pinaster* Ait.). Theoretical and Applied Genetics. 2001;**102**:97-103

[46] Muratova EN. Cytogenetical study on Scots pine (*Pinus sylvestris* L.) in the Central Yakutia. In: Borzan Z, Schlarbaum SE, editors. Cytogenetic Studies of Forest Trees and Shrub Species, Proceedings of the First IUFRO Cytogenetics Working Party S2.04-08 Symposium, September 8–11, 1993. Brijuni National Park Croatia. Zagreb: Faculty of Forestry, University of Zagreb, Zagreb, Croatia; 1997. pp. 157-177

[47] Muratova EN. Studies on nucleolar chromosomes in representatives of Pinaceae Lindl. In: Borzan Z, Schlarbaum SE, editors. Cytogenetic studies of forest trees and shrub species. Proceedings of the First IUFRO Cytogenetics Working Party S2.04-08 Symposium, September 8–11, 1993. Brijuni National Park Croatia. Zagreb: Faculty of Forestry, University of Zagreb, Zagreb, Croatia; 1997. pp. 45-72

[48] Plomion C, Chagné D, Pot S, et al. Pines – Chapter 2. In: Kole C, editor. Genome Mapping and Molecular Breeding in Plants, Volume 7 Forest Trees. Berlin Heidelberg: Springer-Verlag; 2007

[49] Hizume M, Shibata F, Matsusaki Y, et al. Chromosome identification and comparative karyotypic analyses of four *Pinus* species. Theoretical and Applied Genetics. 2002;**105**:491-497

[50] IPCC. Climate change 2022: Impacts, adaptation, and vulnerability. In: Pörtner H-O, Roberts DC, Tignor M, Poloczanska ES, Mintenbeck K, Alegría A, Craig M, Langsdorf S, Löschke S, Möller V, et al., editors. Contribution of Working Group II to the Sixth Assessment Report of the Intergovernmental Panel on Climate Change. Cambridge, UK; New York, NY, USA: Cambridge University Press; 2022. p. 3056

[51] Manso R, Fortin M, Calama R, et al. Modelling seed germination in forest tree species through survival analysis. The *Pinus pinea* L. case study. Forest and Ecology Management. 2013;**289**:515-524

[52] Candel-Pérez D, Linares JC, Viñegla B, et al. Assessing climate-growth relationships under contrasting stands of co-occurring Iberian pines along an altitudinal gradient. Forest and Ecology Management. 2012;**274**:48-57

[53] Carvalho A, Pavia I, Fernandes C, et al. Differential physiological and genetic responses of five European Scots pine provenances to induced water stress. Journal of Plant Physiology. 2017; **215**:100-109. DOI: 10.1016/j.jplph.2017.05.027

[54] Férriz M, Martin-Benito D, Cañellas I, Gea-Izquierdo G. Sensitivity to water stress drives differential decline and mortality dynamics of three co-occurring conifers with different drought tolerance. Forest Ecology and Management. 2021;**486**:118964. DOI: 10.1016/j.foreco.2021.118964

[55] Fernandes C, Gaspar MJ, Pires J, et al. Within and between-tree variation of wood density components in *Pinus*

sylvestris at five sites in Portugal. European Journal of Wood and Wood Products. 2017;**75**(4):511-526. DOI: 10.1007/s00107-016-1130-2

[56] Fernandes C, Gaspar MJ, Pires J, et al. Physical, chemical and mechanical properties of *Pinus sylvestris* wood at five sites in Portugal. iForest – Biogeosciences and Forestry. 2017;**10**(4): 669-679. DOI: 10.3832/ifor2254-010

[57] Dias A, Gaspar MJ, Carvalho A, et al. Within and between-tree variation of wood density components in *Pinus nigra* at six sites in Portugal. Annals of Forest Science. 2018;**75**:58. DOI: 10.1007/s13595-018-0734-6

[58] Dias A, Carvalho A, Silva ME, et al. Physical, chemical and mechanical wood properties of *Pinus nigra* growing in Portugal. Annals of Forest Science. 2020b;**77**:72. DOI: 10.1007/s13595-020-00984-8

[59] Diaci J, Adamič T, Rozman A, et al. Conversion of *Pinus nigra* plantations with natural regeneration in the Slovenian Karst: The importance of intermediate, gradually formed canopy gaps. Forests. 2019;**10**:1136. DOI: 10.3390/f10121136

[60] Scaltsoyiannes A, Rohr R, Panetsos KP, et al. Allozyme frequency distributions in 5 European populations of black pine (*Pinus nigra* Arnold). 1. Estimation of genetic variation within and among populations. 2. Contribution of isozyme analysis to the taxonomic status of the species. Silvae Genetica. 1994;**43**:20-30

[61] Afzal-Rafii Z, Dodd RS. Chloroplast DNA supports a hypothesis of glacial refugia over postglacial recolonization in disjunct populations of black pine (*Pinus nigra*) in Western Europe: Phylogeography of European black pine. Molecular Ecology. 2007;**16**:723-736

[62] Akkemik Ü, Yılmaz H, Oral D, et al. Some changes in taxonomy of pines (*Pinus* L.) native to Turkey. Journal of the Faculty of Forestry, Istanbul University. 2010;**61**:63-78

[63] Rubio Moraga A, Candel Pérez D, Lucas Borja ME, et al. Genetic diversity of *Pinus nigra* Arn. populations in Southern Spain and Northern Morocco revealed by inter-simple sequence repeat profiles. International Journal of Molecular Sciences. 2012;**13**:5645-5658

[64] Gaussen H, Heywood VH, Chater AO. The genus *Pinus* L. In: Tutin TG, Heywood VH, Burgers NA, Valentine DH, Walters SM, Webb DA, editors. Flora Europaea. Vol. 1. Cambridge: Cambridge University Press; 1964. pp. 32-35

[65] Christensen K. Pinaceae, Cupressaceae, Taxaceae, Ephedraceae, Salicaceae, Juglandaceae, Betulaceae, Fagaceae, Ulmaceae Moraceae. In: Strid A, Tan K, editors. Flora Hellenica. Königstein: Koeltz Scientific Books; 1997

[66] Farjon A. World Checklist and Bibliography of Conifers. Kew: Royal Botanic Gardens; 1998

[67] Farjon A. A Handbook of the World's Conifers. Vol. 1. Boston: Brill Leiden; 2010

[68] Price RA, Liston A, Strauss SH. Phylogeny and systematics of *Pinus*. In: Richardson DM, editor. Ecology and Biogeography of *Pinus*. London: Cambridge University Press; 1998. pp. 49-68

[69] Little E, Critchfield W. Subdivisions of the Genus *Pinus* (Pines). Washington: U.S. Department of Agriculture Forest

Services, Miscellaneous Publications; 1969

[70] Louro V. O pinheiro larício: *Pinus nigra* Arnold em Portugal. Lisboa: Direcção Geral do Ordenamento e Gestão Florestal; 1982 (*In Portuguese*)

[71] Kerr G. Natural regeneration of Corsican pine (*Pinus nigra* subsp. laricio) in Great Britain. Forestry. 2000;**73**: 479-488

[72] Lucas Borja ME, Van Stan JT, Plaza-Álvarez PA, et al. Simultaneous estimation of *Pinus nigra* Arn. ssp. *salzmannii* natural regeneration emergence and survival through lifetime analysis. Forest Ecology and Management. 2021;**499**:119613. DOI: 10.1016/j.foreco.2021.119613

[73] Calama R, Conde M, de-Dios-García J, et al. Linking climate, annual growth and competition in a Mediterranean forest: *Pinus pinea* in the Spanish Northern Plateau. Agricultural and Forest Meteorology. 2019;**264**: 309-321

[74] Vázquez-González C, Lopez-Goldar X, Alía R, et al. Genetic variation in resin yield and covariation with tree growth in maritime pine. Forest Ecology and Mangement. 2021;**482**:118843. DOI: 10.1016/j.foreco.2020.118843

[75] Ding X, Li Y, Zhang Y, et al. Genetic analysis and elite tree selection of the main resin components of slash pine. Frontiers in Plant Science. 2023;**14**: 1079952. DOI: 10.3389/fpls.2023.1079952

[76] Lai M, Zhang L, Lei L, et al. Inheritance of resin yield and main resin components in *Pinus elliottii* Engelm. at three locations in southern China. Industrial Crops and Products. 2020;**144**: 112065. DOI: 10.1016/j.indcrop.2019.112065

[77] Lucas-Borja ME, Ahrazem O, Candel-Pérez D, et al. Evaluation of fire recurrence effect on genetic diversity in maritime pine (*Pinus pinaster* ait.) populations using inter-simple sequence repeat profiles. Science of the Total Environment. 2016;**572**:1322-1328

[78] Agee JK. Fire and pine ecosystems. In: Richardson DM, editor. Ecology and Biogeography of *Pinus*. Cambridge: University Press; 1998. pp. 193-218

[79] De las Heras J, Moya D, Vega JA, et al. Post-fire management of serotinous pine forests. In: Moreira F, Arianotsou M, Corona P, De las Heras J, editors. Post-Fire Management and Restoration of Southern European Forests: Managing Forest Ecosystems. Dordrecht: Springer; 2012. pp. 151-170. DOI: 10.1007/978-94-007-2208-8_6

[80] Ribeiro S, Cerveira A, Soares P, Fonseca T. Natural regeneration of maritime pine: A review of the influencing factors and proposals for management. Forests. 2022;**13**:386. DOI: 10.3390/f13030386

Chapter 5

Sexual Dimorphism in Physiological Reactions to Biotope Type (the Case Study in Ground Beetles)

Eugeniy Khomitskiy, Tamara Avtaeva, Shapaat Kushalieva, Alexandr Zamotajlov, Rifgat Shagidullin and Raisa Sukhodolskaya

Abstract

Sexual size dimorphism (SSD) (the difference between males and females in size) is considered to be the physiological reaction to environmental conditions. We estimated body size variation and SSD in two populations of ground beetle *Carabus exaratus*. Beetles were sampled in Krasnodar Province (Russia) in the forb-cereal meadow and beech-hornbeam forest. We processed morphometric measurements in 521 specimen for six linear traits. Beetles from the meadow were slightly larger than from the forest. SSD was pronounced in a half of traits being the highest by elytra length. On a whole, SSD did not differ in the populations studied.

Keywords: sexual size dimorphism, ground beetles, body size variation, physiological reaction, stress

1. Introduction

A lot of studies deal with the relationships between the sizes of different parts or entire bodies of organisms. They investigate how such body size variation affects different morphologies and important functions, that is, physical strength, metabolic rates, and transport of resources [1]. Such size patterns can be studied both within individuals and species and across species of plant or animal. Male and female insects often differ morphometrically, by size [2]. Females are usually larger than males, while males demonstrate lower phenotypic plasticity [3, 4]. Phenotypic plasticity is induced by ecological and environmental variables.

The aim of our study was to top up the knowledge in sexual size dimorphism (SSD) in insects, specifically in ground beetles. The former studies showed SSD variability in these beetles depending on environmental conditions. We chose Caucasian species 0 *Carabus exaratus* Quencel, which was not sufficiently investigated. So we hypothesized that: (i) SSD in it is female-biased; (ii) the beetles from different biotopes will differ in body size; and (iii) SSD by different traits will vary and will be the highest by elytra width.

2. Materials and methods

2.1 Object of investigation

C. exaratus is large (22–28 mm) beetle with elongated body and long legs. It is distributed from foothill steppes till subnival zone. Dominates at foothills in North-Eastern Caucasus and plain parts of Ciscaucasia [5]. Recorded in Rostov and Krasnodar Provinces. Generalist, forest-meadow species, predator with spring-summer breeding.

2.2 Sampling area

We trapped beetles in two biotopes at Krasnodar Province, Tuapse district, Russia. The first biotope was forb-cereal meadow (44.3833°N 38.9667°E, 550 m asl), periodically used for *Zea mays* L. and *Solanum tuberosum* L. cultivation and grazing. The second one was beech-hornbeam forest in Maliy Pshish rever valley (44.1561°N 39.5279°E 520 m asl). Beetles were sampled in 2018 and transported to Laboratory of Biomonitoring Institute for Problems of Ecology and Mineral Wealth Use of Tatarstan Academy of Sciences for measuring. Presented work is the fragment of the investigations in our laboratory. Beginning from 1996, we sampled ground beetles in different regions of Russia and abroad. Beetles are sampled by professional carabidologists for their own purposes and then are posted to our laboratory for the morphometric analysis, because the main aim of our investigation is the clarifying ecogeographical rules (Bergmann, Rensch ets.). So we have to process material in male and female body size variation in latitude and longitude gradients. Thus, samples from the region mentioned (Krasnodar Province) were the part of the material we need to fulfill our aims.

2.3 Morphometric analysis

Nikon D5100 camera was used to get images of beetles. We acted according to the methodology approved in our laboratory, since morphometric analysis requires certain skills and must be carried out by permanent employees. We measured them using a program. The latter was written specifically for the given method of measurement and utilized a distance between manually pointed out elements of photo arrays as the terminal point of measurements and the fiducial scale. Initial codes from the program are available under the free permissive license MIT [6]. Six traits were under investigation: the length and the width of elytra, pronotum, and head. In total, 480 specimens were measured.

2.4 Statistical analysis

Statistical analysis was done using Excel programs. Sexual Size Dimorphism (hereinafter SSD) was calculated as follows: mean value of trait in females/mean value of trait in males-1 [7].

3. Results and discussion

SSD in ground beetles is female-biased [8]. It is common for the vast number of species in invertebrates. The value of SSD is not constant because phenotypic plasticity of body size can vary substantially between the sexes and can cause intraspecific variation [9]. The most interesting opinion on SSD was expressed in Evolutionary Theory of Sex [10]: Males are the vanguard of evolution. They are more variable than females and become very variable in unusual conditions. Purely mathematically, this leads to the reduction of differences between males and females. So in severe environment SSD values increase. On the contrary, in stable environment, without any destabilizing effects, SSD will decrease. Some studies in ground beetles confirmed that suggestion. In mountain populations of ground beetles *Carabus odoratus* Schill. and *Pterostichus montanus* Motsch. SSD highly correlated with population number [11]. It is well known that these conditions lead to resources lack, stress etc. And exactly in those severe conditions SSD in those populations was high. In relation to *C. exaratus*, the highest values in SSD were recorded in severe conditions of high mountains [10]. In our study, the situation was somewhat another. We recorded SSD not by all traits investigated (**Figures 1** and **2**): In elytra length SSD was pronounced in both biotopes, but it was not recorded in elytra width, and was male-biased by pronotum length in the meadow population, and was not recorded in pronotum width and head parameters in the forest population of *C. exaratus*.

Those facts evidenced that environmental conditions were quite stable in area studied. Probably estimated biotopes differed slightly in food resource procuring. Beetles were slightly larger in the meadow, and larva nutrition is considered the main factor that impacts imagoes size. Ground beetles are largely predators feeding on soil micro- and macrofauna representatives—microbes, Formicidae, Silphidae larvae, small species of Staphylinidae, Coccinellidae, Dermestidae, larvae of Lepidoptera, Hemiptera, Thomisidae, Opiliones, and Lumbricidae [12, 13]. Sometimes, carabids consume their "relatives," for example, *Pterostichus melanarius* Ill. fed readily on *Stomis pumicatus* (Panzer, 1795), *Harpalus amplicollis* Mentries, 1848, *Panagaeus bipustulatus* (Fabricius, 1775) [14]. Distribution of ground beetles preys is spatially very diverse. This fact can explain biotope impact on beetle size variation. In arable lands of Europe, the size in ground beetle *Poecilus cupreus* L. in rye differed greatly from the crops. The authors correlated with that fact to the peculiarities of rye cultivation: strong root system and shallow seed embedding. Besides, beetle body

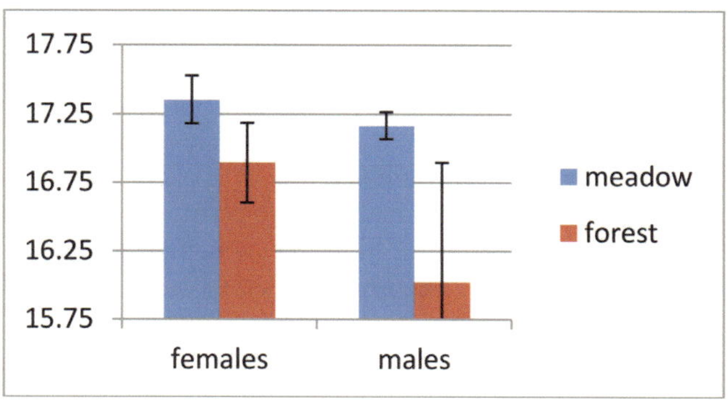

Figure 1.
Elytra length variation in C. exaratus.

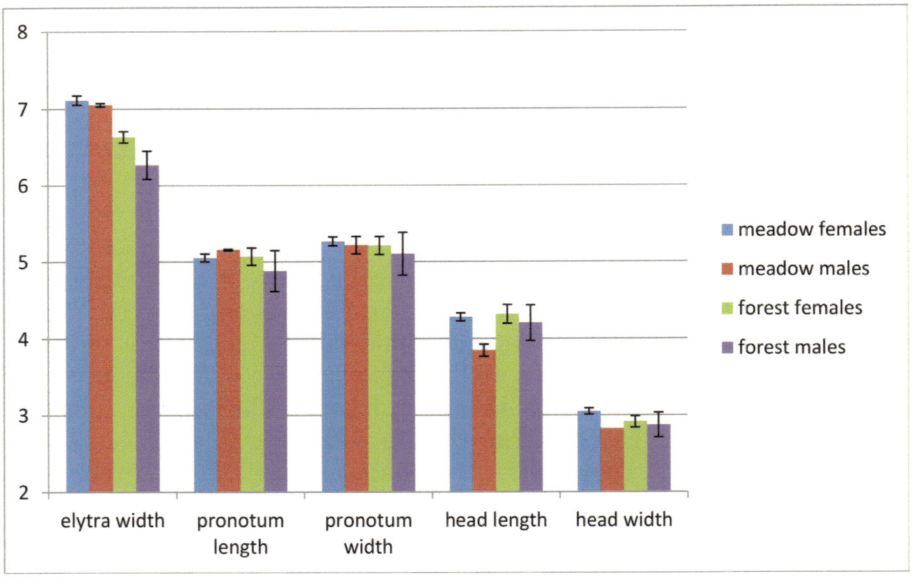

Figure 2.
Traits value variation in C. exaratus.

size and their shape differed in the crops with heavy above-ground parts (carrot, pea, vetch, and oat) and rarefied ones (lawn, meadows) [15]. In *Carabus granulatus* L. beetles were larger in the meadows, but significantly smaller in oak and elm forests if compared with birch ones [16]. Such the differences are based on the different litter decomposition in cited biotopes, where carabid preys dwell [17].

In our case, the beetles in the forest and at the meadow did not differ in size significantly (except their elytra width). This, to our mind, means that the place in area should be taken into account too when exploring body size variation in different populations of ground beetles [18, 19]. For several species of carabids, the distinct curves of size variation were shown in latitude gradient [19] and often males and female response to latitude gradient differed [20, 21]. The plots we explored were situated in nearby locations, so they did not differ significantly in weather conditions and climate in a whole. The latter also affects ground beetle size variation [22, 23].

Besides real size variation the variation of shape should be studied. Our investigated species—*C. exaratus*—showed distinct shape variation in altitude gradient with elongated specimen in low mountains [24].

As for sexual size dimorphism the samples we studied followed the established rule that SSD in carabids is female-biased [25] females being larger than males. On a whole when considering all traits we did not find significant differences in SSD value in the meadow and forest populations (**Figure 3**).

In any way our findings contribute greatly into our knowledge in SSD variation in both carabids and insects. Intra-specific level of investigation allows us to frame mechanisms of body size variation in different habitats and in different sexes. The latter highlight ways of SSD evolution. And it is sufficiently important as SSD is considered to be the indicator of environmental quality.

As mentioned in Material and Methods we are presenting here the fragment of our large study. In previous papers, we put forward the hypothesis that body size variation in ground beetles is genera-specific in latitude gradient: In Carabus beetles, the curve

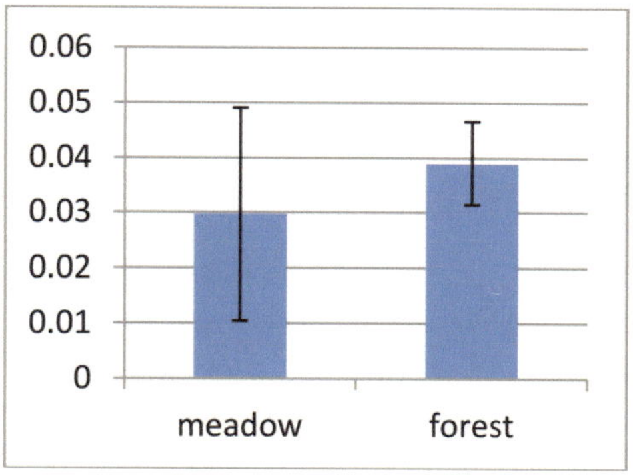

Figure 3.
Sexual size dimorphism (as the mean by all traits investigated) in C. exaratus.

monotonically falls northwards but in Pterostichus ones it is saw-tooth. The latter can be explained by life-cycle changes. As for studied *C. exaratus*, we have the only one region else to be compared with Krasnodar Province—Chechen Republic [10]. The following step of our studies will include comparative analysis of this species samples from different regions and replenishment of our dataset with new data.

Author details

Eugeniy Khomitskiy[1], Tamara Avtaeva[2,3], Shapaat Kushalieva[2], Alexandr Zamotajlov[1], Rifgat Shagidullin[4] and Raisa Sukhodolskaya[4,5,6]*

1 Federal State Budgetary Educational Institution of Higher Education, "I. T. Trubilin Kuban State Agrarian University", Krasnodar, Russia

2 Chechen State Pedagogical University, Grozny, Russia

3 Kh. Ibragimov Complex Institute of the Russian Academy of Sciences, Grozny, Russia

4 Institute for Problems of Ecology and Mineral Wealth Use of Tatarstan Academy of Sciences, Kazan, Russian Federation

5 Kazan State Medical University, Russia

6 Volga-Kama State Reserve, Russia

*Address all correspondence to: sukhodolskayaraisa@gmail.com

References

[1] Zhou X et al. Dynamic allometric scaling of tree biomass and size. Nature Plants. 2021;**7**:42-49

[2] Blanckenhorn W, Fairbairn D, Székely T. Sex, Size and Gender Roles: Evolutionary Studies of Sexual Size Dimorphism. Oxford: Oxford University Press; 2007. 248 p

[3] Stillwell R, Blanckenhorn W, Teder T, Davidowitz G, Fox C. Sex differences in phenotypic plasticity affect variation in sexual size dimorphism in insects: From physiology to evolution. Annual Review of Entomology. 2010;**55**:227-245. DOI: 10.1146/annurevento-112408-085500

[4] Vea I, Wilcox A, Frankino W, Shingleton A. Sex-specific plasticity explains genetic variation in sexual size dimorphism in drosophila. bioRxiv. 2021:448738. DOI: 10.1101/2021.06.16.448738

[5] Aidamirova TA. Carabid assemblages (Coleoptera, Carabidae) at the foothills of the North-Eastern Caucasus. Bulletin of Moscow Nature Researchers. 2011;**116**:20-30

[6] Mukhametnabiev T. Manual Carabid Morphometric Measurement for Method by Sukhodolskaya. Available from: https://github.com/CRTmatrix/-Manual-Carabid-morphometric-measurement-for-method-by-Sukhodolskaya [Accessed: December 1, 2021]

[7] Lovich J, Gibbons J. A review of techniques for quantifying sexual size dimorphism. Growth Development and Aging. 1992;**56**:269-281

[8] Sukhodolskaya R, Avtaeva T, Saveliev A, Vavilov D. Sexual size dimorphism in ground beetle *Carabus cumanus* Fischer von Waldheim, 1823 (Coleoptera, Carabidae) and its variation in different traits. Baltic Journal of Coleopterology. 2019;**19**(1):89-100

[9] Teder T, Tammaru T. Sexual size dimorphism within species increases with body size in insects. Oikos. 2005;**108**(2):321-334

[10] Geodakyan VA. Two Sexes. Why? The Evolutionary Theory of Sex. Dichron; 2012

[11] Sukhodolskaya R, Ananina T, Avtaeva T, Saveliev A. Sexual size dimorphism in ground beetles and its variation in altitude gradient. Advances in Medicine and Biology. 2021;**191**:1-62. Available from: https://novapublishers.com/?s=Advances+in+Medicine+and+Biology.+Volume+191&post_type=product

[12] Korolev OV, Brygadyrenko VV. Comparative analysis of *Pterostichus melanarius* (Coleoptera, Carabidae) trophic preferences in different condition of laboratory keeping. Optima Prots Ecosystem. 2012;**6**:178-190

[13] Korolev OV, Brygadyrenko VV. Trophic relations of *Pterostichus melanarius* (Coleoptera, Carabidae) with dominant species of invertebrates in forest ecosystems of steppe Dnieper region. Visnyk of Dnipropetrovsk University. Biology, Ecology. 2021;**20**:48-54

[14] Brygadyrenko VV, Korolev OV. Peculiarities of trophic spectrum of *Pterostichus melanarius* (Coleoptera: Carabidae) in laboratory conditions. Visnyk Bilocerk. Derzhavnogo Agrar. Univ. 2006;**43**:67-71

[15] Sukhodolskaya R, Saveliev A. Crop impact on body size variation in carabid beetle *Poecilus cupreus* Linnaeus (Coleoptera, Carabidae). Ukrainska Entomofaunistyka. 2016;7(3):84

[16] Sukhodolskaya R, Saveliev A. Impact of environmental factors on the body shape variation and sexual shape dimorphism in *Carabus granulatus* L. (Coleoptera: Carabidae). Zoological Systematics. 2017;**42**(1):71-89

[17] Beugnon R, Eisenhauer N, Bruelheide H, Davrinche A, Du J, Haider S, et al. Tree diversity effects on litter decomposition are mediated by litterfall and microbial processes. Oikos. 2023;**2023**:e09751. DOI: 10.1111/oik.097512

[18] Mukhametnabiev T, Sukhodolskaya R, Vorobyova I, Antsiferov A, Ukhova N. Influence of geographic location in area and dominant forest forming species on body shape of ground beetle *Pterostichus oblongopunctatus* Fabricius, 1787 (Coleoptera: Carabidae) in taiga-broadleaf gradient in Russia. Russian Journal of Applied Ecology N. 2020;**1**:3-12

[19] Sukhodolskaya R, Saveliev A. Body size variation in ground beetles (Coleoptera: Carabidae) in latitude gradient. Periodicum Biologorum. 2016;**118**(3):273-278

[20] Luzyanin S, Saveliev A, Ukhova N, Vorobyova I, Solodovnikov I, Anciferov A, et al. Modeling sexual differences of body size variation in ground beetles in geographical gradients: A case study of *Pterostichus melanarius* (Illiger, 1798) (Coleoptera, Carabidae). Life. 2022;**12**:112. DOI: 10.3390/life12010112

[21] Sukhodolskaya R, Saveliev A, Ukhova N, Vorobyova I, Solodovnikov I, Anciferov A, et al. Modeling sexual differences of body size variation in ground beetles in geographical gradient (the case study in *Pterostichus oblongpunctatus* Fabricius, 1787). GSC Biological and Pharmaceutical Sciences. 2020;**13**(03):149-161. DOI: 10.30574/gscbps.2020.13.3.0388

[22] Luzyanin S, Gordienko T, Saveliev A, Ukhova N, Vorobeva I, Solodovnikov I, et al. Impact of climatic factors on sexual size dimorphism in ground beetle *Pterostichus melanarius* (Illiger, 1798) (Coleoptera, Carabidae). Ecologica Montenegrina. 2022;**58**:1-13

[23] Ananina T, Saveliev A, Shagidullun R, Gordienko T, Sukhodolskaya R. Climatic factors can differently affect body size in closely related species (the case study in ground beetles). Russian Journal of Applied Ecology N. 2023;**3**:4-11

[24] BenítezH, SukhodolskayaR, AvtaevaT, Escobar-Suárez S, Órdenes-Claveria R, Laroze D, et al. Quantifying elevational effect on the geometric body shape of Russian beetle *Carabus exaratus* (Coleoptera: Carabidae). Zoologischer Anzeiger. 2023;**302**:30-36

[25] Sukhodolskaya R, Saveliev A, Muhammetnabiev T. Sexual dimorphism of insects and conditions of its manifestation. Research Journal of Pharmaceutical, Biological and Chemical Sciences. 2016;7(2):1992-2001

Section 4

Genetic Structures and Characteristics

Chapter 6

Population Genetic Structure of Marine Fishes

Fidelina Gonzalez, Patricio Barria, Francisco Ponce and Sergio Mora

Abstract

Research on population genetics structure of marine fishes is increasing because of new technology based on DNA sequencing. This knowledge is necessary for management and conservation of natural population in marine environment. The aim of the chapter is to discuss about how genetic population structure get from DNA, allows us to know about dynamic of life history of species of Teleosts (Actinopterigii) and Chondrichthyes (Elasmobranchii). The analysis is based on taxonomic point of view. We hope to contribute to apply the new advances to management of natural population of fishes and marine wildlife.

Keywords: population genetic structure, life history, Actinopterigii, Elasmobranchii, marine environment

1. Introduction

Marine fishes are highly dynamic organisms that live in an also dynamic environment. Temperature and salinity variations combine to determine the density of seawater. Density governs the vertical movements of water, and density-driven circulation. Life of fishes depends on availability of food, and refugia to complete their lifetime in this environment. Tiger sharks (*Galeorcerdo cuvier*, Galeocerdonidae) moved along about 6–8 km. Thresher sharks (*Alopias vulpinus*, Alopidae) move, at least 400–1900 m, while learning of adult threshers improves the ability to perform directed walks. Blacktip reef sharks (*Carcharhinus melanopterus*, Carcharhinidae) showed high site fidelity and no evidence of oriented movements at large scales and small home ranges [1].

Population structure of marine fishes, otherwise, is determined by genetics or genomics information. Some aspects of this could be expressed in movements of individuals in their lifetime such as philopatry, natal philopatry (return to birth location) or spawning philopatry (return to a previous spawning site used by adults), should be included in stocks management. Natal philopatry improves the probability of offspring to find the best nursery habitats. Aggregation behavior in male and female leopard sharks, *Triakis semifasciata* (Triakidae), strong seasonal philopatry, with 50.0% of females and 60.0% of males moving every year to their aggregation sites [2]. An experiment realized in shark has shown that *Triakis semifasciata* can use their olfaction for navigation. Tracks of control sharks ended 62.6% closer to shore. Tracks of anosmic sharks (nares occluded with cotton wool) follow tortuous paths, ended 37.2% closer to shore. Detection and discrimination of odors is the function of olfaction, and for navigation [3]. The largest olfactory bulb is found in migratory coastal-pelagic species such as *Carcharodon carcharias* (Lamnidae), *Galeocerdo cuvier*, and *Prionace glauca* (Carcharhinidae) [4]. Gardiner *et al.* [5] have shown that olfaction participates in homing by juvenile *Carcharhinus limbatus* within a shallow bay. Evidence for olfaction-mediated homing in fish larvae, whereas one species showed panmixis, another species showed strong homing. Thus, larvae smelling allow us to choose currents to return to natal home. Therefore, reef populations can develop genetic differences that might lead to reproductive isolation [6], operating mostly in nearshore environments [7].

2. Methodology

To develop this work, we search the information using Pubmed and Taxonomy from NCBI, Google Scholar Databases, and species with barcode in Bold System (http://www.boldsystems.org), and review the references related to DNA obtained from marine fishes using Scholar Google.

3. History life cycles of marine fishes

Fish life histories are necessary to understand how to preserve fish populations for future generations and to make practical regulations for fisheries management (for example, not taking certain species during spawning season or releasing undersized fish). Overexploitation, habitat destruction, pollution, and the uncontrolled introduction of exotic species for mass fish production have led to a decrease in the diversity of species in marine environment. In addition, climate change, which leads to changes in atmospheric temperatures that affect circulation patterns in aquatic environments, results in changes in salinity and dissolved oxygen and affects reproduction cycles of species.

Habitat fragmentation is one of the greatest threats to global biodiversity [8], preventing migratory organisms from completing their life cycles [9] and restricting dispersal and gene flow between populations [10]. The global decline of 76% in abundance of migratory fish over the past 50 years is due to fragmentation of habitat [11].

About 35,000 species of fish have been quantified worldwide. Marine fishes comprise about 60% of fish species in the world. Marine fish have three different areas in their life cycle. The first is for reproduction, the second for juvenile rearing and the third for adult feeding. This implies displacements or movements of different magnitude (migration) among the members of a species forming a panmictic population. Knowing the dynamics of the individuals of a particular population means considering these factors of each species. The interaction between environmental forces and dispersal characteristics is responsible for the population structure patterns of marine fishes. Sequential hermaphroditism defined as individuals first maturing as one sex and later in life transforming into the other sex, modifying population structure patterns. Sex-changing fishes are more spatially structured than gonochoristic (separate sexes) species because of skewed sex ratio. F_{ST} index, test the hypothesis that sex-changing species that are more prone to genetic drift are more genetically structured than gonochoristic species [12].

According to Pla et al. [13] described that functional hermaphroditism has been confirmed in the following orders: Anguilliformes, Aulopiformes, Centrarchiformes, Gobiiformes, Perciformes, Scorpaeniformes, Stomiiformes, and Synbrachiformes. The five families with the higher number of hermaphrodite species were Serranidae (91 spp.), Labridae (63 spp.), Gobiidae (41 spp.), Sparidae (40 spp.), and Scaridae (23 spp.). Sequential hermaphroditism predominated over simultaneous hermaphroditism (305 vs. 65 spp.), 244 are protogyny species, 39 protandric species, and 22 species are bi-directional sequential hermaphroditism. In sequential (sex-changing) hermaphrodite fishes, the sex ratio is usually biased toward the first sex, while reproductive success increases considerably after sex change. Estimated *Ne*, effective population size (*Ne*) in natural populations, the number of individuals that effectively participate in producing the next generation, from two ecologically similar species from the east coast of South Africa, there were *Ne* estimates significantly lower in the protogynous species and no evidence of genetic structuring or significant variation in genetic diversity in the study area [14].

Management and conservation can be greatly informed by considering explicitly how environmental factors influence population genetic structure. A new approach to characterize population connectivity at small spatial scales to conservation and fisheries management for high gene flow species is given by recent advances in oceanographic approaches for larval dispersal [15].

To evaluate genetic connectivity among three subpopulations of *Bothus robinsi* (Pleuronectiformes) Morales-Pulido *et al.* [16] sampled larvae from the Bay of Campeche and the eastern Gulf of Mexico and adults on Florida's continental shelf to test if the larvae tend to mix or display collective dispersal, compare genetic information between larvae and adults. Using ddRadseq (used for SNP discovery and genotyping which is a variation on the RAD sequencing protocol) RAD means restriction site associated DNA markers to genotype 1034 single-nucleotide polymorphic sites (SNPs) from *B. robinsi* larvae, identified morphologically and by DNA barcoding. A long pelagic larval duration of *B. robinsi* and ocean dynamics to transport and mix larvae from all shelf area is enough to produce connectivity, indicating a panmictic population.

Predictions on population-level genetic connectivity and the directionality of effective dispersal, as example, in two seahorses (*Hippocampus*) and three pipefishes (*Syngnathus*), of family Syngnathidae, using genome-wide single-nucleotide polymorphism (SNP) data, it found that follow the predominant ocean circulation patterns in the Gulf of Mexico and northwestern Atlantic [17]. High levels of gene flow and directionality are associated with the pelagic macro-algae *Sargassum* spp.

3.1 DNA molecular markers and main taxonomic group of fishes

To study the population genetics of marine fishes should be necessary to identify the species as the first step. A molecular technique to identify species is DNA barcoding, utilizing mitochondrial gene cytochrome oxidase I. By DNA barcoding is possible to identify species and categorize species for conservation. Mitochondrial Cytochrome c Oxidase subunit I (COI) is a gene of 650 base pairs that codes a protein, important in the cellular respiration and is considered a universal barcode for animals, because of its fast mutation rate, and maternal inheritance, it can be accessed from the DNA libraries data record. DNA barcoding gives information on cryptic species [18]. The benefits of barcoding fishes include species identification, range expansion for known species, and taxonomic identifications where traditional methods cannot be applied. Specimens with divergent barcode sequences have been confirmed by taxonomic analysis as new species [19]. However, DNA barcoding, with the support of traditional taxonomy has the capability to identify species complexes within populations [20].

More than 25,000 species have already been DNA barcoded. 85% of a total of 1415 species sharks and rays of 12 orders are barcoded (**Table 1**). 95% of a total of 26,133 species grouped in 70 orders of Actinopterigii; a total of 458107sequences stored by orders, and 24,761 species are barcoded, in Bold System. The barcode sequence is a useful tool to identify fish species from any fish, fillet, fin, egg or larva can be matched against reference sequences using barcode of life data (BOLD) system [21] (http://www.boldsystems.org).

MitoFish is a comprehensive and standardized fish mitochondrial genome database of fish mitochondrial genomes (mitogenomes) that includes powerful and precise de novo annotations for mitogenome sequences [22] (http://mitofish.aori.u-tokyo.ac.jp/). Mitogenomic sequence data have resolved fish phylogenies and identifying new fish species [23]. Mitofish database contains 137 mitogenome of species of Elasmobranchii (**Table 1**), and 2316 species of Actinopterigii.

3.2 Cryptic species

Cryptic species require special attention in conservation planning, especially for endangered species complex. Cryptic species are biological entities based on morphological similarities classified as single species, but they possess distinct genetic

Order	Sequences BOLD system	Total species	Species with barcode	Mitofish
Carcharhiniformes	13,136	326	267	55
Heterodontiformes	133	9	7	2
Hexanchiformes	254	6	6	5
Lamniformes	1675	16	15	14
Myliobatiformes	6929	438	376	34
Orectolobiformes	1115	38	34	9
Pristiformes	1572	90	78	0
Pristiophoriformes	68	6	5	1
Rajiformes	6353	261	225	40
Squaliformes	4313	144	118	0
Squatiniformes	488	26	24	8
Torpediniformes	645	55	49	6
Echinorhiniformes	—	—	—	0
Rhinopristiformes	—	—	—	13
Total	36,681	1415	1204	137

Last column includes information about species sequenced mitocondrial DNA in Mitofish.

Table 1.
List of orders of Elasmobranchii, total sequences stored by orders, total species, and species with barcode in Bold system.

lineages and different species might require different conservation strategies [24]. Some species collected from complex ecosystems or different geographical locations showed higher intraspecific genetic distance values due to sympatric or allopatric speciation process. Sympatric speciation, the formation of species in the absence of geographical barriers diverse lineages through selective divergence in Carangoidei [25, 26]. In allopatric speciation, due to geographical barrier that prevent gene flow between populations, same species could be developed into different genetic lineages and develop genetic differences with little morphological change [27].

Analysis of DNA barcoded 2276 specimens belonging to 668 coral reef fish species from Indian and Pacific oceans it proved to be highly effective for the discovery of provisional cryptic diversity [28]. Of the 141 species sampled on each side of the Indo Malaysian Philippine Archipielag, 62 species presented no spatial structure whereas 67 exhibited divergent lineages.

Luciobrotula is a small genus of benthopelagic fishes occurring at depths ranging from 115 to 2300 m in the Atlantic, Indian, and Pacific Oceans, with six species, of the family Ophidiidae (Ophidiformes). *Luciobrotula bartschi* is the only known species in the West Pacific. Wong *et al.* [29] described *L. polylepis* as a new species, previously identified as *L. bartschi*. The genetic distance between the two species is 13.8% on average at the COI locus.

DNA barcoding is a powerful tool to study biodiversity, to increase the knowledge of marine biodiversity. The Southeast Pacific comprises two Large Marine Ecosystems, the Pacific Central American Coastal and the Humboldt Current System show a low representation of barcoded species, more abundant for the Humboldt Current System than the Pacific Central American Coastal. Ramirez *et al.* [30] found for Actinopterygii (30.27%) and Elasmobranchii (24.71%) high level of barcode incongruences for fish using Barcode Index Number (BIN).

In Nigerian fish samples collected from coastal waters, Bolaji *et al.* [31] find four species identified using COI, for the first time, for Nigerian marine environment: *Caranx fischeri* (Carangidae), *Pseudotolithus senegallus* (Sciaenidae), *Lagocephalus guentheri* (Tetraodontidae), and *Sphyraena ensis* (Sphyraenidae).

Using 1314 ultraconserved elements defined as highly conserved regions within the genome that are shared among evolutionarily distant taxa [32], the phylogenomic analysis of Syngnathidae reveals a high biodiversity inside the family [33]. The genus *Corythoichthys* (Syngnathidae) comprises a group of taxonomically complex, widely distributed in the IndoPacific region. To understand the genetic diversity, distribution, and 'species groups' within the genus *Corythoichthys*, a phylogenetic analysis carried out using 52 COI sequences. The analysis confirmed 13 species considered as a single taxon in the IndoPacific realm [34].

The genetic identity of marine species is kept through physiological or behavioral mechanisms that prevent gene flow and hybridization: ecological niche, environmental tolerance, mate recognition, differences in spawning time, reproductive migration, and gamete compatibility may be involved in marine fish species definition [35]. DNA barcoding is a tool to discover cryptic species in Mugilidae necessary for conservation and management of fisheries. Morphological examination of 40 mullets reveals 6 known species but the DNA barcode-based species identify eight operational taxonomic units (OTUs) belonging to five genera (*Crenimugil*, *Ellochelon*, *Mugil*, *Osteomugil*, and *Planiliza*) [36] in Pakistan, a biogeographic area where nearly no mullet species were genetically characterized. A sampling of 245 specimens studied from the Indo-Australian Archipelago found that belonged to five different genera. Some of these are known species, cryptic diversity was found for three of these species, and eight (30%) probably new species [37]. The population structure of *Mugil cephalus* was investigated across 18 sampling sites in the NE Atlantic Ocean, Mediterranean, and Black Seas, using an 857 bp fragment of mitochondrial (mtDNA) cytochrome b, and 7 nuclear loci. The low nucleotide diversity of mtDNA marker was very low (0.6% divergence) because it is a conservative sequence, indicating a single clade over the entire area. Between the samples from the Black Sea and the samples from other basins, it finds significant mtDNA genetic differentiation observed ($\Phi_{ST} = 0.17$; $p = 0.029$). Significant genetic differentiation and isolation-by-distance in *M. cephalus* find based on nuclear loci analysis [38].

DNA barcoding using mitochondrial DNA sequences may offer an effective approach to identifying cryptic species and characterizing genetic diversities of deep-sea fish species [39]. DNA barcoding of mesopelagic and demersal fish collected around Japan and southern Taiwan. High intraspecific genetic differentiation showed based on COI sequences of samples of *Chimaera phantasma* (Chimaeridae), *Harpadon microchir* (Synodontidae), and *Pyramodon ventralis* (Carapidae). Moreover, for 19 widespread deep-sea fish species, suggested a high level of genetic differentiation for the northwestern Pacific Ocean, that suggest many cryptic species or regional populations not described yet, in deep-sea fishes [40].

To identify development stages of lifetime for some species, the eggs of 15 species collected at 40 stations distributed across and around the Gulf of Mexico. The cluster analysis identified eight groups that yielded percomorph eggs, a neritic community, an oceanic community, and a transitional community in the shelf break. The coastal-pelagic species would have mixed neritic–oceanic distributions. Of the 1144 eggs barcoded, 709 (62%) identified 62 species belong to 42 families, and 20 taxa identified to genus or subfamily level [41].

3.3 Mitochondrial DNA control region. Subpopulation.

The lack of a known repair mechanism for mutations that arise during replication of mitochondrial DNA allows higher rate of evolution than nuclear genome. The initiation of replication of mitochondrial DNA occurs in the control region (about 1 KB). Control region of mitochondrial DNA is a good marker for analyzing population connectivity, defining conservation units, identifying migration routes of fishes, using in many phylogenetic and population genetic studies because of high copy number, and its maternal and haploid mode of inheritance [42].

Flounders provided evidence of a genetic structure between the sampling sites from Southwest Atlantic coast using a fragment of the mitochondrial DNA of the control region and seven microsatellite loci. *Paralichthys orbignyanus* displayed a lower to moderate contemporary genetic structure among all samples except between Lagoa dos Patos (Brazil) to San Matias Gulf (Argentina), explained probably by limited gene flow associated with a consistent larvae retention in all sampling sites, congruent with the prevailing littoral drift [43].

The mitochondrial DNA control region of 91 individuals from six populations Japanese flounder *Paralichthys olivaceus* (Paralichthyidae), which is endemic to the Northwest Pacific. The analysis of 233 sequences, from four populations, showed high nucleotide diversity (0.032 ± 0.016) and haplotype diversity (0.996 ± 0.001). The geographical distribution and genetic population structure of *P. olivaceus* is explained by ocean currents and climate change. The peripheral population in the East China Sea had the relatively lowest genetic diversity and highest differentiation [44].

Biological parameters, geopolitical boundaries, and the existence of fisheries management units do not match genetic populations for marine fisheries management. The contribution of genetic and genomic data allows to define management units. *Pagellus bogaraveo* (Sparidae) is a sparid in the northeast Atlantic, with three stock units. The northern stock collapsed (1975–1985). The Mediterranean and Azores are two different units. A significant population structure found for the entire sample area (AMOVA: $F_{ST} = 0.052$, $p < 0{,}001$) attributed to the individuals from the Azores, which showed significant differences with the remaining five sites. Mitochondrial control region corroborates the results obtained using nuclear markers (i.e. microsatellites) [45]. Consistent results get from mitochondrial and nuclear markers [46]. The differentiation of the Azorean population is common to other species of the Northeast Atlantic, *Labrus bergylta* (Labridae) [47], *Aphanopus carbo* (Aphanopodinae) [48], *Coryphoblennius galerita* (Blenniidae) [49] and *Diplodus vulgaris* (Sparidae) [50].

Lumpfish (*Cyclopterus lumpus*, Cyclopteridae), a transatlantic marine fish, show high dispersal, and gene-flow, displaying large population sizes. Nevertheless, this species may have a natal homing behavior and local populations with adaptive differences because of the observed high levels of population structuring throughout the Atlantic. Analysis using SNP methodological strategies by Jansson et al. [51] identified genetic evidence of local population substructures [51].

3.4 Conservation of marine fishes

Climate change and anthropogenic activity affect marine biodiversity in the Persian Arabian Gulf. Assessment following the International Union for Conservation of Nature (IUCN) Red List methodology, of 471 species of marine bony fishes in the Gulf, informed about the threatened marine bony fishes. Based on available data all species, threats include 47% of marine bony fishes related to fisheries and harvesting, and 32% of marine fish species related to coastal development and loss of habitat, acute in nearshore areas where spatial analyses indicated high species richness [52].

Fisheries resource management in the Persian Arabian Gulf has been inadequate but only 22 Gulf fish species have been listed as threatened or Near Threatened due to impacts from overexploitation. Classified as a primary threat 120 species with large-scale exploitation, 105 species listed as Least Concern including species with a small maximum size, short lifespan, and high reproductive output; for example, six of the eight clupeids commonly found in fisheries were listed as Least Concern [52].

3.5 New technologies and future development

Marine ecosystems are currently subjected to a range of exploitation rates. Management actions have controlled reductions in exploitation rates in some regions, but a fraction of stocks collapsed. Local context could define the selection the best management tools. A combination of traditional approaches (catch quotas, community management) coupled with strategically placed fishing closures, changing selective fishing gear, holds much promise for restoring marine fisheries and ecosystems [53, 54].

Recently advances in environmental DNA (eDNA) originates from cellular material shed by organisms in aquatic or terrestrial environments is revolutionizing fish biodiversity monitoring. To improve the sensitivity and reliability of fish eDNA analyses are recommended the available universal primers used to amplify barcoding sequences from fish eDNA and combined use of different gene marker such as pair of primers of conservative sequences of following genes, such as12SrDNA, 16SrDNA, Cytochrome b, Cytochrome oxidase 1, and databases [55]. The first step in improving the accuracy and efficiency of eDNA detection is optimized a technical procedure for eDNA-based detection with high specificity to identify the COI gene [56].

Based on metabarcoding of the 16S mitochondrial gene for fish and the 18S nuclear gene for macroinvertebrates of Manly Lagoon, a polluted estuary in eastern Australia DiBattista et al. [57] identified seasonal differences in fish and macroinvertebrate community composition, as well as species richness to examine temporal and spatial changes in the aquatic community, which correlated (in some cases) with the environmental parameters of sea surface temperature and freshwater input. Several known migratory fish species contributed significantly to the overall patterns observed (fish assemblages shifting in response to environmental drivers in polluted estuaries).

A new tool for knowing the spatio-temporal population structure of fish species and diversity is the massive sequencing of genetic markers, such as the use of single-nucleotide polymorphisms (SNP) obtained by DNA sequencing associated with the restriction site by RADseq. They find in the genome and define a thousand loci [58–60].

The overexploitation of shallow-water fish stocks has led to increasing fishing pressure on deep-sea species. Genetic structure of *Pagellus bogaraveo* find no variation in *cyt-b* and low genetic variation in *D-loop* sequences, nucleotide diversity was extremely low ($p = 0.002$) for mitochondrial DNA in fish and haplotype diversities ($h = 0.5919$), effective population size must have been reduced to very low numbers, indicate that *P. bogaraveo* may have undergone a severe bottleneck in the past, as other Atlantic species of fish, possibly originated from the last glaciation [61].

Hydrographic or topographic barriers prevent dispersal of adults and/or larvae in deep-sea fish species between populations at regional and oceanographic scales. Analysis of population structure using D-loop and microsatellite indicates significant genetic differentiation between populations at a regional level. Using SNPs obtained by RADseq, Francisco *et al.* [46, 62] identified different populations of *Pagellus bogaraveo* in the Atlantic Ocean. The results of this investigation suggest the need to integrate this type of information in stock management.

A marine fish in the North Pacific, *Gadus macrocephalus* (Gadidae), experienced shifts in biomass and distribution linked to climate change [62]. During the spawning season of fish collected around the Korean Peninsula using DNA (RAD) sequencing and individual assignment defined small and isolated peripheral populations, often remnants of glacial refugia in high gene flow marine species. Populations relatively small of *Gadus macrocephalus* on Korean coasts were highly differentiated ($F_{ST} = 0.025–0.042$). Data suggest asymmetrical dispersal and gene flow, potentially involving adaptive alleles, between peripheral populations inhabiting markedly different thermal regimes. Fisher *et al.* (2022) [63] suggest the conservation value of peripheral populations [64, 65]. A loss of genetic diversity that includes adaptive variation important for response to environmental change [65]. Adaptive alleles in small peripheral populations of *Gadus macrocephalus* demonstrate contemporary migration from warm- to cold-adapted populations, Peripheral populations should therefore be conserved and maintained as an important genetic resource.

Eleginops maclovinus (Eleginopsidae), is a marine protandric fish, male in the early stages of life live [66, 67], in estuarine habitats, and female in the later stages, when it is pelagic, live near the coast. They tend to live at the mouths of rivers, adults are not morphologically hydrodynamics, spawning and developing in this environment [68]. The dynamic and heterogeneous landscape in Patagonia, can harbor an important but cryptic genetic population structure using restriction site-associated DNA (RAD) sequencing. In five locations sampled, distributed across a salinity cline from Northern Patagonia, Canales *et al.* [69] identified a spatial structuration with gene flow and spatial selection by environmental association. Neutral loci gave two genetic group meanwhile adaptive loci gave three genetic groups. The effective population sizes ranged from 572 to 14,454 and were influenced by locality.

Analysis of specimens of *Boreogadus saida* (Gadidae) collected in the Canadian Arctic Archipelago would set potential barriers to gene flow between the Canada West group and Canada East. Differentiation patterns can be explored by the application of single-nucleotide polymorphism markers (SNPs) [70]. Detected population differentiation of *B. saida* over circumpolar and regional scales. Patterns of gene flow can be influence by geography, river discharge, prevailing currents, bathymetry, water properties, and prey availability. Direction of gene flow has important implications for expansion of southern alleles, prevailing currents may be promoting northward range expansions [71].

The high trophic level oceanodromous species undertake migrations of significant distances across the oceans to feed, or to breed, and have wide geographic distributions, such as tuna. Therefore, these species live both within the exclusive economic zones and outside these zones. Pelagic species live primarily in the open ocean and spend part of their life cycle in nearshore waters [72]. Microsatellites and SNP arrays do not identify population structure and fish populations from different ocean basins. However, stock genomics research using RAD sequencing can give genetic differentiation in *Thunnus albacares* (Scombridae) [73–75] *Thunnus alalunga* (Scombridae) [76, 77], *Acanthocybium solandri* (Scombridae) [78], and *Kajikia audax* (Istiophoridae) [79].

4. Conclusion

The new advances in DNA technology and bioinformatics allow to increase the knowledge, as show in this chapter, both fish biology and population genetics, therefore, to contribute to improving the management of natural population of fishes and marine wildlife.

Conflict of interest

The authors declare no conflict of interest.

Author details

Fidelina Gonzalez[1*], Patricio Barria[2], Francisco Ponce[3] and Sergio Mora[4]

1 Faculty of Biological Sciences, Department Cell Biology, University of Concepción, Concepción, Chile

2 Instituto de Fomento Pesquero, Valparaíso, Chile

3 Retiree, Concepción, Chile

4 Instituto de Fomento Pesquero, Talcahuano, Chile

*Address all correspondence to: fgonzale@udec.cl

References

[1] Papastamatiou YP, Cartamil DP, Lowe CG, Meyer CG, Wetherbee BM, Holland K. Scales of orientation directed walks and movement path structure in sharks. Journal of Animal Ecology. 2011;**80**(4):864-874. DOI: 10.1111/j.1365-2656.2011.01815.x

[2] Nosal AP, Caillat A, Kisfaludy EK, Royer MA, Wegner NC. Aggregation behavior and seasonal philopatry in male and female leopard sharks *Triakis semifasciata* along the open coast of southern California, USA. Marine Ecology Progress Series. 2014;**499**:157-175. DOI: 10.3354/meps10632

[3] Jacobs LF. From chemotaxis to the cognitive map: The function of olfaction. Proceedings of the National Academy of Sciences. 2012;**109**(supplement_1):10693-10700. DOI: 10.1073/pnas.1201880109

[4] Yopak KE, Lisney TJ, Collin SP. Not all sharks are "swimming noses": Variation in olfactory bulb size in cartilaginous fishes. Brain Structure and Function. 2015;**220**:1127-1143. DOI: 10.1007/s00429-014-0705-0

[5] Gardiner JM, Whitney NM, Hueter RE. Smells like home: The role of olfactory cues in the homing behavior of blacktip sharks, *Carcharhinus limbatus*. Integrative and Comparative Biology. 2015;**55**(3):495-506. DOI: 10.1093/icb/icv087

[6] Gerlach G, Atema J, Kingsford MJ, Black KP, Miller-Sims V. Smelling home can prevent dispersal of reef fish larvae. Proceedings of the National Academy of Sciences. 2007;**104**(3):858-863. DOI: 10.1073/pnas.0606777104

[7] Nosal AP, Chao Y, Farrara JD, Chai F, Hastings PA. Olfaction contributes to pelagic navigation in a coastal shark. PLoS One. 2016;**11**(1):e0143758. DOI: 10.1371/journal.pone.0143758

[8] Wilson MC, Chen XY, Corlett RT, Didham RK, Ding P, Holt RD, et al. Habitat fragmentation and biodiversity conservation: Key findings and future challenges. Landscape Ecology. 2016;**31**:219-227. DOI: 10.1007/s10980-015-0312-3

[9] Lucas MC, Bubb DH, Jang MH, Ha K, Masters JE. Availability of and access to critical habitats in regulated rivers: Effects of low-head barriers on threatened lampreys. Freshwater Biology. 2009;**54**(3):621-634. DOI: 10.1111/j.1365-2427.2008.02136.x

[10] Wilkes MA, Webb JA, Pompeu PS, Silva LG, Vowles AS, Baker CF, et al. Not just a migration problem: Metapopulations, habitat shifts, and gene flow are also important for fishway science and management. River Research and Applications. 2019;**35**(10):1688-1696. DOI: 10.1002/rra.3320

[11] Kerr JR, Tummers JS, Benson T, Lucas MC, Kemp PS. Modelling fine scale route choice of upstream migrating fish as they approach an instream structure. Ecological Modelling. 2023;**478**:110210. DOI: 10.1016/j.ecolmodel.2022.110210

[12] Chopelet J, Waples RS, Mariani S. Sex change and the genetic structure of marine fish populations. Fish and Fisheries. 2009;**10**(3):329-343. DOI: 10.1111/j.1467-2979.2009.00329.x

[13] Pla S, Maynou F, Piferrer F. Hermaphroditism in fish: Incidence, distribution and associations with abiotic environmental factors. Reviews in Fish Biology and Fisheries.

2021;**31**(4):935-955. DOI: 10.1007/s11160-021-09681-9

[14] Coscia I, Chopelet J, Waples RS, Mann BQ, Mariani S. Sex change and effective population size: Implications for population genetic studies in marine fish. Heredity. 2016;**117**(4):251-258. DOI: 10.1038/hdy.2016.50

[15] White C, Selkoe KA, Watson J, Siegel DA, Zacherl DC, Toonen RJ. Ocean currents help explain population genetic structure. Proceedings of the Royal Society B: Biological Sciences. 2010;**277**(1688):1685-1694. DOI: 10.1098/rspb.2009.2214

[16] Morales-Pulido JM, Munguia-Vega A, Jiménez-Rosenberg SPA, Rocha-Olivares A, Galindo-Sánchez CE. Genetic connectivity in Twospot flounder (Bothus robinsi) across the Gulf of Mexico, inferred with single nucleotide polymorphisms from larvae and adults. Frontiers in Marine Science. 2023;**10**:1070483. DOI: 10.3389/fmars.2023.1070483

[17] Bertola LD, Boehm JT, Putman NF, Xue AT, Robinson JD, Harris S, et al. Asymmetrical gene flow in five co-distributed syngnathids explained by ocean currents and rafting propensity. Proceedings of the Royal Society B. 1926;**2020**(287):20200657. DOI: 10.1098/rspb.2020.0657

[18] Thiriet PD, Di Franco A, Cheminee A, Guidetti P, Bianchimani O, Basthard-Bogain S, et al. Abundance and diversity of crypto-and necto-benthic coastal fish are higher in marine forests than in structurally less complex macroalgal assemblages. PLoS One. 2016;**11**(10):e0164121. DOI: 10.1371/journal.pone.0164121

[19] Ward RD, Hanner R, Hebert PD. The campaign to DNA barcode all fishes, FISH-BOL. Journal of Fish Biology. 2009;**74**(2):329-356. DOI: 10.1111/j.1095-8649.2008.02080.x

[20] Iwatsuki Y, Tanaka F, Allen GR. *Lutjanus xanthopinnis*, a new species of snapper (Pisces: Lutjanidae) from the indo-West Pacific, with a redescription of Lutjanus madras (Valenciennes 1831). Journal of the Ocean Science Foundation. 2015;**17**:22-42

[21] Bold Systems [Internet]. [Accessed: May 30, 2023]. Available from: https://www.boldsystems.org/

[22] MitoFish [Internet]. [Accessed: June 18, 2023]. Available from: https://mitofish.aori.u-tokyo.ac.jp/

[23] Iwasaki W, Fukunaga T, Isagozawa R, Yamada K, Maeda Y, Satoh TP, et al. MitoFish and MitoAnnotator: A mitochondrial genome database of fish with an accurate and automatic annotation pipeline. Molecular Biology and Evolution. 2013;**30**(11):2531-2540. DOI: 10.1093/molbev/mst141

[24] Bickford D, Lohman DJ, Sodhi NS, Ng PK, Meier R, Winker K, et al. Cryptic species as a window on diversity and conservation. Trends in Ecology & Evolution. 2007;**22**(3):148-155. DOI: 10.1111/mec.14486

[25] Glass JR, Harrington RC, Cowman PF, Faircloth BC, Near TJ. Widespread sympatry in a speciesrich clade of marine fishes (Carangoidei). bioRxiv. 2022-09. 2022. DOI: 10.1101/2022.09.26.509594

[26] Kimura S, Takeuchi S, Yadome T. Generic revision of the species formerly belonging to the genus Carangoides and its related genera (Carangiformes: Carangidae). Ichthyological Research. 2022;**69**(4):433-487. DOI: 10.1007/s10228-021-00850-1

[27] Leray M, Beldade R, Holbrook SJ, Schmitt RJ, Planes S, Bernardi G. Allopatric divergence and speciation in coral reef fish: The three-spot *Dascyllus*, *Dascyllus trimaculatus*, species complex. Evolution. 2010;**64**(5):1218-1230. DOI: 10.1111/j.1558-5646.2009.00917.x

[28] Hubert N, Meyer CP, Bruggemann HJ, Guerin F, Komeno RJ, Espiau B, et al. Cryptic diversity in indo-Pacific coral-reef fishes revealed by DNA-barcoding provides new support to the Centre-of-overlap hypothesis. PLoS One. 2012;**7**(3):e28987. DOI: 10.1371/journal.pone.0028987

[29] Wong MK, Lee MY, Chen WJ. Integrative taxonomy reveals a rare and new cusk-eel species of *Luciobrotula* (Teleostei, Ophidiidae) from the Solomon Sea. West Pacific. European Journal of Taxonomy. 2021;**750**:52-69. DOI: 10.5852/ejt.2021.750.1361

[30] Ramirez JL, Rosas-Puchuri U, Cañedo RM, Alfaro-Shigueto J, Ayon P, Zelada-Mazmela E, et al. DNA barcoding in the Southeast Pacific marine realm: Low coverage and geographic representation despite high diversity. PLoS One. 2020;**15**(12):e0244323. DOI: 10.1371/journal.pone.0244323

[31] Bolaji DA, Lawal-Are AO, Kuton MP. DNA barcoding and misidentification of some marine fish species in Nigerian industrial trawl fishery. Scientific African. 2023;**20**:e01662. DOI: 10.1016/j.sciaf.2023.e01662

[32] Bejerano G, Pheasant M, Makunin I, Stephen S, Kent WJ, Mattick JS, et al. Ultraconserved elements in the human genome. Science. 2004;**304**(5675):1321-1325. DOI: 10.1126/science.1098119

[33] Stiller J, Short G, Hamilton H, Saarman N, Longo S, Wainwright P, et al. Phylogenomic analysis of Syngnathidae reveals novel relationships, origins of endemic diversity and variable diversification rates. BMC Biology. 2022;**20**(1):1-21. DOI: 10.1186/s12915-022-01271-w

[34] Shalu K, Thomas L, Ramvilas G, Shabeena KS, Philip S, Sureshkumar S, et al. DNA barcodes for the pipefish genus *Corythoichthys* (Actinopterygii: Syngnathiformes) from the Indian Ocean provide insights into cryptic diversity. Journal of Fish Biology. 2013;**102**(3):680-688. DOI: 10.1111/jfb.15300

[35] Palumbi SR. Genetic divergence, reproductive isolation, and marine speciation. Annual Review of Ecology and Systematics. 1994;**25**(1):547-572

[36] Hasan A, Siddiqui PJA, Amir SA, Durand JD. DNA barcoding of mullets (family Mugilidae) from Pakistan reveals surprisingly high number of unknown candidate species. Diversity. 2021;**13**(6):232. DOI: 10.3390/d13060232

[37] Delrieu-Trottin E, Durand JD, Limmon G, Sukmono T, Kadarusman SHY, et al. Biodiversity inventory of the grey mullets (Actinopterygii: Mugilidae) of the indo-Australian archipelago through the iterative use of DNA-based species delimitation and specimen assignment methods. Evolutionary Applications. 2020;**13**(6):1451-1467. DOI: 10.1111/eva.12926

[38] Cossu P, Mura L, Scarpa F, Lai T, Sanna D, Azzena I, et al. Genetic patterns in *Mugil cephalus* and implications for fisheries and aquaculture management. Scientific Reports. 2021;**11**(1):2887. DOI: 10.1038/s41598-021-82515-7

[39] Teramura A, Koeda K, Matsuo A, Sato MP, Senou H, Ho HC, et al. Assessing the effectiveness of DNA barcoding for exploring hidden genetic diversity in deep-sea fishes.

Marine Ecology Progress Series. 2022;**2022**(701):83-98. DOI: 10.3354/meps14193

[40] Teramura A, Koeda K, Senou H, Ho HC, Kikuchi K, Hirase S. DNA barcoding of deep-sea fishes from the northwestern Pacific Ocean: a resource for identifying hidden genetic diversity in deep-sea fishes. Authorea. 2021. Available online: https://www.authorea.com [Accessed: June 18, 2023]

[41] Burrows M, Browning JS, Breitbart M, Murawski SA, Peebles EB. DNA barcoding reveals clear delineation between spawning sites for neritic versus oceanic fishes in the Gulf of Mexico. Fisheries Oceanography. 2019;**28**(2):228-239. DOI: 10.1111/fog.12404

[42] Verma R, Singh M, Kumar S. Unraveling the limits of mitochondrial control region to estimate the fine scale population genetic differentiation in anadromous fish *Tenualosa ilisha*. Scientifica. 2016;**2016**:2035240. DOI: 10.1155/2016/2035240

[43] Fainburg L, Sabadin D, Díaz de Astarloa JM, Fernández IP. Population structure of mud flounder *Paralichthys orbignyanus* from the South-Western Atlantic Ocean. Journal of Fish Biology. 2023;**102**(2):455-464. DOI: 10.1111/jfb.15271

[44] Sun CH, Yang F, Huang Q, Zeng XS, Zhang YN, Li S, et al. Genetic population structure and demographic history of the endemic fish *Paralichthys olivaceus* of the Northwest Pacific Ocean. Ecology and Evolution. 2022;**12**(11):e9506. DOI: 10.1002/ece3.9506

[45] Piñera JA, Blanco G, Vázquez E, Sánchez JA. Genetic diversity of blackspot seabream (*Pagellus bogaraveo*) populations off Spanish coasts: A preliminary study. Marine Biology. 2007;**151**:2153-2158. DOI: 10.1007/s00227-007-0665-5

[46] Robalo JI, Farias I, Francisco SM, Avellaneda K, Castilho R, Figueiredo I. Genetic population structure of the blackspot seabream (*Pagellus bogaraveo*): Contribution of mtDNA control region to fisheries management. Mitochondrial DNA Part A. 2021;**32**(4):115-119. DOI: 10.1080/24701394.2021.1882445

[47] Almada F, Francisco SM, Lima CS, Fitz Gerald R, Mirimin L, Villegas-Ríos D, et al. Historical gene flow constraints in a northeastern Atlantic fish: Phylogeography of the ballan wrasse *Labrus bergylta* across its distribution range. Royal Society open. Science. 2017;**4**(2):160773. DOI: 10.1098/rsos.160773

[48] Stefanni S, Knutsen H. Phylogeography and demographic history of the deep-sea fish *Aphanopus carbo* (Lowe, 1839) in the NE Atlantic: Vicariance followed by secondary contact or speciation? Molecular Phylogenetics and Evolution. 2007;**42**(1):38-46. DOI: 10.1016/j

[49] Francisco SM, Almada VC, Faria C, Velasco EM, Robalo JI. Phylogeographic pattern and glacial refugia of a rocky shore species with limited dispersal capability: The case of Montagu's blenny (*Coryphoblennius galerita*, Blenniidae). Marine Biology. 2014;**161**:2509-2520. DOI: 10.1007/s00227-014-2523-6

[50] Stefanni S, Castilho R, Sala-Bozano M, Robalo JI, Francisco SM, Santos RS, et al. Establishment of a coastal fish in the Azores: Recent colonisation or sudden expansion of an ancient relict population? Heredity. 2015;**115**(6):527-537. DOI: 10.1038/hdy.2015.55

[51] Jansson E, Faust E, Bekkevold D, Quintela M, Durif C, Halvorsen KT,

et al. Global, regional, and cryptic population structure in a high gene-flow transatlantic fish. PLoS One. 2023;**18**(3):e0283351. DOI: 10.1371/journal.pone.0283351

[52] Buchanan JR, Ralph GM, Krupp F, Harwell H, Abdallah M, Abdulqader E, et al. Regional extinction risks for marine bony fishes occurring in the Persian/Arabian gulf. Biological Conservation. 2019;**2019**(230):10-19. DOI: 10.1016/j.biocon.2018.11.027

[53] Worm B, Hilborn R, Baum JK, Branch TA, Collie JS, Costello C, et al. Rebuilding global fisheries. Science. 2009;**325**(5940):578-585. DOI: 10.1126/science.1173146

[54] Memarzadeh M, Britten GL, Worm B, Boettiger C. Rebuilding global fisheries under uncertainty. Proceedings of the National Academy of Sciences. 2019;**116**(32):15985-15990. DOI: 10.1073/pnas.1902657116

[55] Xiong F, Shu L, Zeng H, Gan X, He S, Peng Z. Methodology for fish biodiversity monitoring with environmental DNA metabarcoding: The primers, databases and bioinformatic pipelines. Water Biology and Security. 2022;**1**(1):100007. DOI: 10.1016/j.watbs.2022.100007

[56] Sun M, Guo Y, Zhao N, Zhang S, Pei K, Qin C. Fish eDNA detection and its technical optimization: A case study of *Acanthopagrus latus*. Marine Environmental Research. 2022;**176**:105588. DOI: 10.1016/j.marenvres.2022.105588

[57] DiBattista JD, Fowler AM, Riley IJ, Reader S, Hay A, Parkinson K, et al. The use of environmental DNA to monitor impacted coastal estuaries. Marine Pollution Bulletin. 2022;**181**:113860. DOI: 10.1016/j.marpolbul.2022.113860

[58] Andrews KR, Good JM, Miller MR, Luikart G, Hohenlohe PA. Harnessing the power of RADseq for ecological and evolutionary genomics. Nature Reviews Genetics. 2016;**17**(2):81-92. DOI: 10.1038/nrg.2015.28

[59] Benestan L. Population genomics applied to fishery management and conservation. In: Oleksiak M, Rajora O, editors. Population Genomics: Marine Organisms. Cham: Springer; 2020. pp. 399-421

[60] Helyar SJ, Hemmer-Hansen J, Bekkevold D, Taylor MI, Ogden R, Limborg MT, et al. Application of SNPs for population genetics of nonmodel organisms: New opportunities and challenges. Molecular Ecology Resources. 2011;**11**:123-136. DOI: 10.1111/j.1755-0998.2010.02943.x

[61] Stockley B, Menezes G, Pinho MR, Rogers AD. Genetic population structure in the black-spot sea bream (*Pagellus bogaraveo* Brünnich, 1768) from the NE Atlantic. Marine Biology. 2005;**146**:793-804. DOI: 10.1007/s00227-004-1479-3

[62] Francisco SM, Robalo JI. Time matters: Genetic composition and evaluation of effective population size in temperate coastal fish species. PeerJ. 2020;**8**:e9098. DOI: 10.7717/peerj.9098

[63] Fisher MC, Helser TE, Kang S, Gwak W, Canino MF, Hauser L. Genetic structure and dispersal in peripheral populations of a marine fish (Pacific cod, *Gadus macrocephalus*) and their importance for adaptation to climate change. Ecology and Evolution. 2022;**12**(1):e8474. DOI: 10.1002/ece3.8474

[64] Hampe A, Petit RJ. Conserving biodiversity under climate change: The rear edge matters. Ecology Letters. 2005;**8**(5):461-467. DOI: 10.1111/j.1461-0248.2005.00739.x

[65] Nadeau CP, Urban MC. Eco-evolution on the edge during climate change. Ecography. 2019;**42**(7):1280-1297. DOI: 10.1111/ecog.04404

[66] Calvo J, Morriconi E, Rae GA, San Roman NA. (1992). Evidence of protandry in a subantarctic notothenid, *Eleginops maclovinus* (Cuv. & Val., 1830) from the Beagle Channel, Argentina. Journal of Fish Biology. 1992;**40**(2):157-164. DOI: 10.1111/j.1095-8649.1992.tb02563.x

[67] Brickle P, Laptikhovsky V, Arkhipkin A. Reproductive strategy of a primitive temperate notothenioid Eleginops maclovinus. Journal of Fish Biology. 2005;**66**(4):1044-1059. DOI: 10.1111/j.1095-8649.2005.00663.x

[68] Guzmán L, Campodónico I. Algunos aspectos de la biología de *Eleginops maclovinus* (Cuv. y Val.) 1830, con especial referencia a su morfometría, caracteres merísticos y alimentación. In Anales del Instituto de la Patagonia. Instituto de la Patagonia. Punta Arenas. Región de Magallanes. Chile. 1973. Guzman_ Anales_1973_vol4_p.343-371.pdf

[69] Canales-Aguirre CB, Larson WA, McKinney GJ, Claure CE, Rocha JD, Ceballos SG, et al. Neutral and adaptive loci reveal fine-scale population structure in *Eleginops maclovinus* from North Patagonia. Ecology and Evolution. 2022;**12**(10):e9343. DOI: 10.1002/ece3.9343

[70] Glover KA, Hansen MM, Lien S, Als TD, Høyheim B, Skaala O. A comparison of SNP and STR loci for delineating population structure and performing individual genetic assignment. BMC Genetics. 2010;**11**:1-12. DOI: 10.1186/1471-2156-11-2

[71] Nelson RJ, Bouchard C, Fortier L, Majewski AR, Reist JD, Præbel K, et al. Circumpolar genetic population structure of polar cod, *Boreogadus saida*. Polar Biology. 2020;**43**:951-961. DOI: 10.1007/s00300-020-02660-z

[72] Popova E, Vousden D, Sauer WH, Mohammed EY, Allain V, Downey-Breedt N, et al. Ecological connectivity between the areas beyond national jurisdiction and coastal waters: Safeguarding interests of coastal communities in developing countries. Marine Policy. 2019;**104**:90-102. DOI: 10.1016/j.marpol.2019.02.050

[73] Grewe PM, Feutry P, Hill PL, Gunasekera RM, Schaefer KM, Itano DG, et al. Evidence of discrete yellowfin tuna (*Thunnus albacares*) populations demands rethink of management for this globally important resource. Scientific Reports. 2015;**5**(1):1-9. DOI: 10.1038/srep16916

[74] Pecoraro C, Babbucci M, Franch R, Rico C, Papetti C, Chassot E, et al. The population genomics of yellowfin tuna (*Thunnus albacares*) at global geographic scale challenges current stock delineation. Scientific Reports. 2018;**8**(1):13890. DOI: 10.1038/s41598-018-32331-3

[75] Moore BR, Bell JD, Evans K, Farley J, Grewe PM, Hampton J, et al. Defining the stock structures of key commercial tunas in the Pacific Ocean I: Current knowledge and main uncertainties. Fisheries Research. 2020;**230**:105525. DOI: 10.1016/j.fishres.2020.105525

[76] Anderson G, Hampton J, Smith N, Rico C. (2019). Indications of strong adaptive population genetic structure in albacore tuna (*Thunnus alalunga*) in the southwest and Central Pacific Ocean. Ecology and Evolution. 2019;**9**(18):10354-10364. DOI: 10.1002/ece3.5554

[77] Vaux F, Bohn S, Hyde JR, O'Malley KG. Adaptive markers distinguish north

and South Pacific albacore amid low population differentiation. Evolutionary Applications. 2021;**14**(5):1343-1364. DOI: 10.1111/eva.13202

[78] Haro-Bilbao I, Riginos C, Baldwin JD, Zischke M, Tibbetts IR, Thia JA. Global connections with some genomic differentiation occur between indo-Pacific and Atlantic Ocean wahoo, a large circumtropical pelagic fish. Journal of Biogeography. 2021;**48**(8):2053-2067. DOI: 10.1111/jbi.14135

[79] Mamoozadeh NR, Graves JE, McDowell JR. Genome-wide SNPs resolve spatiotemporal patterns of connectivity within striped marlin (*Kajikia audax*), a broadly distributed and highly migratory pelagic species. Evolutionary Applications. 2020;**13**(4):677-698. DOI: 10.1111/eva 12892

Chapter 7

Founder Effect: Breeding a Dog for the Elderly Gentleman Reveals an Animal Model of a Human Genetic Disorder

Felicia Ikolo, Sabyasachi Maity, Robert Finn, Atoum Abdullah, Alireza Tajik, Jessie M. Cameron and Mary C. Maj

Abstract

Animal models of genetic disorders that have risen due to selective breeding can be used as a valuable model to teach the basic concepts of population genetics. The Clumber Spaniel is a breed of dog created in the mid-1700s by the 4th Duc du Noailles. He selectively bred this dog for the elderly gentleman. This sleepy-looking breed survives today, though 1% suffer from severe exercise intolerance due to an autosomal-recessive founder mutation in the pyruvate dehydrogenase phosphatase 1 (PDP1) gene. PDP1 deficiency was long suspected to be a human metabolic disorder and described at the molecular level in 2005 by Robinson and coworkers. The Robinson group later identified a founder mutation within the PDP1 gene of the Clumber spaniel. This case clearly illustrates how a detrimental mutant allele in a small population, when selecting for phenotype, can persist in the progeny of that group. In this review, we discuss the origin of the "Founder Effect" theory and present an example of how a bottleneck that occurred during the selective breeding of the Clumber spaniel over 250 years ago led to the current genetic status of the breed. Today, genotyping can help reduce the incidence of PDP1 in the Clumber breed.

Keywords: founder effect, population genetics, PDP1 deficiency, DNM1 deficiency, Clumber spaniel

1. Introduction

Healthy genetic diversity of populations, in humans and all other animals, is subject to fluctuations in population size. In a large population, random mating will occur, and allelic changes that are not compatible with life, or of low fitness, will be removed from the gene pool. If a population becomes somehow limited by the environment, natural selection may influence allelic changes to either increase or disappear. Charles Darwin brought his idea of natural selection forth with the publication of his book "On the Origin of Species". Here he defined that natural selection is a process whereby the population of an organism will adapt and change to suit an environment and that offspring tend to resemble their parents [1]. Notably, mutation is the source of allelic variability and beneficial mutations increase fitness, whereas deleterious mutations reduce fitness. A limited population, or founder population due to some type of bottleneck or separation from the larger parent population, may suffer a loss of genetic variation and thus be closed off to genetic diversity. In a small and limited population, deleterious mutations may accumulate. The Wright-Fisher Model, named for the early pioneers of theoretical population genetics, describes allelic frequencies of a single gene assuming random mating, nonoverlapping generations, and no selection or mutation. This model also considers variations in population sizes and how a sudden reduction in the size may distort allelic frequency. The distortion may lead to allelic variants which were found in extremely low frequency within the general population yet relatively common in isolated groups. In a population with a finite number, a deleterious allele can fix, and thus reduce its fitness. In the absence of outcrossing, the fitness of the group may continue to reduce, up to and including extinction, under the load of the deleterious alleles. There are many examples of inherited human diseases where a pathogenic variant is found at a higher frequency in particular groups or in populations that have an elevated level of consanguinity.

In animals, a phenomenon called the "popular sire" describes the reproductive dominance of specific animals, increasing the likelihood of later matings between related animals. Applying selective pressure based on the phenotypic attributes of the popular sire within a small population of animals markedly restricts genetic diversity. This becomes problematic when there is an increase in the incidence of harmless recessive alleles, or carriers, leading to an increased potential for progeny to inherit the recessive allele in a homozygous fashion, thus resulting in an autosomal recessive disorder. Corrective measures are currently being put in place by kennel clubs around the world such as pedigree analysis, coefficient of inbreeding calculations, and the development of single nucleotide polymorphism (SNP) arrays are currently being used to manage the diversity of breeds and control heritable disorders. Autosomal recessive disorders, which are known to a specific breed can be tested to identify carriers and reduce the numbers of affected pups born. The Clumber spaniel is one of the oldest of all British spaniel breeds, with over a 250-year-old history of joining the hunt with many of Europe's crowned heads and aristocracy beginning in the Georgian era. It is not a common breed, but better known to sporting gundog enthusiasts, being bred originally for work during the hunt rather than for appearance at show. The breed was named after Clumber Park, originally part of the Sherwood Forest and home to the Dukes of Newcastle in Nottinghamshire, England. First arriving in England between 1763 and 1768, the dog was selectively bred in the kennel of the 2nd Duke of Newcastle to work the undergrowth of the Sherwood Forest during the hunt, to be strong, slow, and steady during the gunfire, and have an exceptional nose for the game [2].

When a large population is suddenly reduced, for a variety of reasons, the resulting population will experience a reduction in the genetic variability known as Bottleneck and Founder Effects. Selective crossbreeding of purebred animals will increase the frequency of rare mutations, or alleles, which may cause diseases similar to the Founder Effect. Events that cause a group of humans to become isolated from a larger population, in time, will similarly increase the frequency of rare mutations. For canines and humans alike, autosomal recessive disorders require progeny to inherit two mutated alleles before the disease is expressed (homozygosity). Here, we review the bottleneck events that took place during the development of the Clumber spaniel breed. We describe how selective breeding for traits important for gun hunting, such as slow and steady with a specific nose, has led to the increased frequency of mutations leading to extreme exercise intolerance in the homozygous state. The particular mutation of the autosomal recessive disorder pyruvate dehydrogenase deficiency 1 (PDP1) was first identified by us in humans and determined to be disease-causing through protein complementation assay [3]. Using the same assay conditions on cell cultures obtained from Clumber spaniels, we were able to restore pyruvate dehydrogenase activity. Molecular analysis identified the disease-causing mutation in the PDP1 gene. A genotyping test was developed, and over 100 Clumber spaniel dogs were analyzed. It was determined that the carrier frequency of the disease-causing allele was 20% of all dogs tested [4]. Our historical investigation showing over 250 years of selective breeding of a hunting dog for the elderly gentleman led us to the current molecular diagnosis, as an example of a Founder Effect, similar to that seen in humans. What was once theorized by the pioneers of population genetics can now be proven by modern-day investigation. This review combines historical descriptions of the Founder Effect as it can currently be seen by genotype analysis.

2. Mendel's law and the founder effect

In 1865, Gregor Johann Mendel presented his results of artificial fertilization to achieve color variation with ornamental plants and this work that proposed the laws of inheritance was published the next year [5]. A generation later, when scientists had a better understanding of cells and genetic material, Mendel's work was independently rediscovered by three botanists. In the early years of 1900, Hugo DeVries, Carl Correns, and Erich von Tschermak each confirmed Mendel's laws of inheritance, as reviewed by Wilkie [6]. The first human disorder identified to follow Mendelian inheritance was published in 1902 by Archibald Garrod who described alkaptonuria as being inherited according to Mendelian rules. Garrod wrote of his investigation of a family, where two of five children were secreting homogentisic acid in their urine, which is pathognomonic for a diagnosis of alkaptonuria. Alkaptonuria was thought to be a peculiar disorder at the time for two reasons. First, homogentisic acid could be observed in patient's urine from approximately 60 hours after birth and could be followed for life. Homogentisic acid causes urine to darken after exposure to air. Second, alkaptonuria did not appear to cause any detriment to the health of those with the condition but was thought to be an alternative mode of metabolism. It is now known that while the onset of clinical phenotype can be delayed in life, oxidative forms of homogentisate deposited in sclerae, cartilage, and skin can lead to early osteoarthritis and cardiac valve disease. Garrod made inquiries to several other investigators who were following alkaptonuric patients at the time, and they kindly shared their data for publication. After compiling the data for 11 families, it was found that 60% of families of children with alkaptonuria were born to first-cousin parents [7]. Garrod surmised that alkaptonuria was a congenital disorder that followed Mendelian recessive inheritance, where the rare recessive trait passed through many generations before the union of two recessive traits manifest in one zygote, as was described by Bateson in the same year [8]. The peculiar nature of alkaptonuria and the level of consanguinity in most affected families studied, allowed these early investigators to calculate the incidence of disease and postulate the mode of inheritance at a most opportunistic time when Mendel's laws of inheritance were rediscovered.

At the time when Garrod published his findings on the incidence of alkaptonuria, and into the 1920s, the statistical study of Mendelian laws in a large population was gathering interest. Working independently, Hardy and Weinburg described similar theories that in a sexually reproducing population, the frequency of two alleles will come to equilibrium in one generation and remain unchanged provided that mating is random, the population is large and no additions to the population from outside occurs [9, 10]. The Hardy-Weinberg principle describes the relationship between allelic and genotypic frequencies in a diploid organism from one generation to another. Evolution can alter this equilibrium to include the influence of mutation, recombination, selection, and isolation. The relative frequencies of alleles in large populations can be influenced by factors that disturb equilibrium such as mutation, migration, and Darwinian selection was also considered, along with the concept of the effective population size (*Ne*). *Ne* is a mathematical representation of genetic diversity within an idealized population in comparison to a population of interest [11, 12]. In 1954, Ernst Mayr theorized that if a group of persons, which he classified as "founders", are somehow isolated from their original large, or ancestral, population, they would abruptly undergo reduced levels of genetic variability leading to the accumulation of inbreeding within this new founder population [13]. Population geneticists have developed mathematical models to estimate how allele frequencies in a large, ancestral population can be expected to change with a sudden reduction in the population and lead to a founder effect [13]. The alleles of the founder population, with the accumulation of inbreeding, will level off to a new stable allele frequency [14]. If an unfavorable allele is in the first generation of a founder population, and this unfavorable allele has a fitness between 0.5 and 1, the allele will increase in numbers while the population doubles in size [14, 15]. Experiments have been performed in the laboratory to demonstrate rapid speciation employing an insect model to demonstrate, as recently reviewed by Haudry, Laurent, and Kapun [16].

Human population genetics typically defines the founder effect as a loss in genetic variation that can arise when a new population is derived from a small number of individuals that were segregated from a larger group. The average heterozygosity seen per locus will initially decline in the smaller, reduced group initially. Over time, new mutations will increase as the population rises. Scientists have used this theory to explain why some inherited disorders can be found at a higher frequency in a defined population than is seen in the general population. There is empirical evidence that suggests an individual in a small population that has been isolated by geography, religion, or culture, can carry an alteration to their DNA. This person is referred to as the common ancestor. After many generations, as the population expands, this variation will be seen at high frequency in the group. If the DNA variant is not disease-causing, it can be considered a polymorphism that is found at high frequency in this segregated group. If the variant is disease-causing for an autosomal recessive disorder, we will see an increase in the number of persons expressing the disorder with each successive generation, and this supports the concept of the founder effect. The founder effect has also been seen to occur naturally in animal populations. However, in many cases, it is due to humans who wish to enrich specific traits in a species of animals. Founder effect is one of the causes of the concentration of genes for specific diseases in purebred (genetic isolates) dogs [17]. Breeders of animal stock actively participate in the phenomena of heredity in real-time.

3. The Clumber spaniel

3.1 The breed

An interesting story describing the breeding of desirable traits in animals is that of the Clumber spaniel. This breed of canine is of particular fascination to the authors of this chapter as our work in the Robinson lab in 2005 on a human autosomal recessive disorder affecting pyruvate metabolism, led to the diagnosis of the same metabolic disorder in the Clumber spaniel [3, 4]. There are a few theories regarding the origin of this breed, and each is entwined in romantic, historical folklore. The most common, and rather intriguing theory begins with the aristocracy of eighteenth century France and the Noailles family. Due to the French Revolution, most of the Noailles family documents were destroyed along with many family members. Any papers that survived the Revolution were destroyed in a fire at the Louvre in May 1871 [2]. The lore suggests that in the early 1760s, Louis de Noailles (the future 4th Duc de Noailles), invited his good friend Henry Pelham-Clinton (the future 2nd Duke of Newcastle), to go hunting on his estate in France. Henry Pelham-Clinton was extremely impressed with Louis de Noailles' hunting spaniels, particularly for their prowess in flushing game. Louis de Noailles gifted his breeding stock of Clumbers to Henry Pelham-Clinton, who brought the dogs to his Clumber Park estate in Sherwood Forest, Nottinghamshire, England in 1764. Together with the assistance of William Mansell, his gamekeeper, they further developed this breed for the "retired gentleman" or "elderly gentlemen's hunting companion." This breed of dog would have likely been lost forever in the chaos of the French Revolution as the French stock was wiped out [18]. The first record of Clumbers in England is seen in the 1788 painting by Sir Francis Wheatly as shown in **Figure 1**.

Figure 1.
Sir Francis Wheatley's painting titled "The Return from Shooting" depicts the Duke of Newcastle on the hunt with Colonel Litchfield, his gamekeeper William Mansell, and four white with orange/lemon marked Clumber spaniels, confirming that the breed was likely in England before the of the French revolution which began on or about 1789 and ended in 1799. This picture is in the public domain and the original is in the possession of the Sheffield gallery in Sheffield, England. https://commons.wikimedia.org/w/index.php?curid=8708898

The Clumber spaniel began as a small kennel of pure-bred dogs. No genetic diversity was introduced for approximately 30 years after the breed was brought to Clumber Park and under the direction of William Mansell [18]. True to the initial selection criteria of traits, the breed is not as energetic as other hunting spaniels. The Clumbers seem to require only about an hour of activity daily. The Clumber physique is long and low, typically heavier than other Spaniels, with a white coat that can be easily seen by hunters. They continue to be developed in the United Kingdom as working gun dogs, where modest size, lack of exaggeration, good eyes, good hips, good temperament, responsiveness, athleticism, and working aptitudes are the important qualities [2]. Their personality has been described by the American Kennel Club as mellow, gentlemanly, amiable, dignified but always amusing [19]. Selective breeding of this low-energy dog has resulted in approximately 20% of current live-born Clumber spaniels being carriers for a metabolic deficiency, which results in severe exercise intolerance with lactic acidosis [4]. The metabolic deficiency in the Clumber spaniel was first reported in 1979 by Herrtage and Houlton, who described a dog that collapsed and showed extreme lactic acidosis after moderate exercise [20]. Järvinen and Sankari reported that a 3-year-old Clumber spaniel had a history of exercise intolerance since puppyhood. During exercise, the dog would quickly tire, lay down, and could not rise until resting for up to 1 hour. Post-exercise blood lactate levels were very elevated. The dog died suddenly while investigations were being performed and a defect in pyruvate dehydrogenase activity was suspected [21].

3.2 Etiology of lactic acidosis

Lactic acid is a normal product of metabolism and is utilized mainly in the liver and kidney, which express the requisite enzymes for processing, in a pathway called gluconeogenesis. During exercise, skeletal muscle produces circulating lactate which typically normalizes in time. Lactic acidosis is also a common finding in disease states when there is an increased production accompanied by decreased clearance. Metabolic lactic acidosis occurs when there is a defect in the aerobic metabolism of pyruvate. Pyruvate is naturally generated as the final product of glycolysis. Pyruvate is then used, through a series of enzymatic steps, to generate ATP. ATP is the primary energy molecule that is used by the cell for activities such as muscle contraction, synthesis of macromolecules, and maintenance of the cellular membrane. The normal fate of pyruvate is first to enter the mitochondria where it is converted to acetyl CoA by the pyruvate dehydrogenase complex. Second, acetyl CoA feeds into the citric acid cycle which generates products that drive a series of reactions of the electron transport chain to finally produce ATP. Deficits in any of the enzyme systems of these pathways will result in the accumulation of pyruvate, which in turn is reduced to lactate, thus causing lactic acidosis. The clinical features of lactic acidosis include muscle weakness, confusion, increased heart rate, metabolic acidosis with an elevated anion gap, compensatory hyperventilation due to metabolic acidosis, and in extreme cases may result in coma.

4. Methods to identify the PDP1 mutation in Clumber spaniels

4.1 Development of PDP1 assay in human tissue

Pyruvate dehydrogenase complex (PDH) plays a pivotal role in metabolism and is highly regulated by the concentration of its substrates and products. It is also regulated by phosphorylation and dephosphorylation events, by kinases and phosphatases, respectively. In 1975, Robinson and Sherwood performed biochemical assays for PDH activity in tissue samples isolated from a young patient with metabolic acidosis as a result of elevated blood lactate and pyruvate [22]. In summary, the assay for PDH activity was performed in three parts. The first assesses the activity of the native PDH complex in tissue homogenate. Next, the homogenate is incubated with ATP before assaying PDH activity, to activate kinase phosphorylation which inhibits the PDH complex. Lastly, the homogenate is incubated with calcium before assaying PDH activity, to activate the phosphatase which switches on PDH activity. Their results showed that the patient's PDH complex could not be activated by calcium and proposed that the deficit was a result of defective pyruvate dehydrogenase phosphatase deficiency. After that publication, many physicians around the world sent tissue samples to the Robinson laboratory for the diagnosis of patients with congenital lactic acidosis. In 2005, the fibroblasts of two brothers with congenital lactic acidosis were sent to the Robinson lab for testing, as reported by us [3]. In summary, native PDH activity was 25% of control which could not be restored by preincubation with calcium. Maj and coworkers of the Robinson lab generated recombinant pyruvate dehydrogenase phosphatase isoform 1 (PDP1) and isoform 2 (PDP2) proteins. Patient PDH activity was restored to normal after incubation with recombinant PDP1, suggesting a defect not in the PDH complex itself but in the regulatory PDP1 protein. Mutational analysis was performed and a homozygous, pathogenic mutation in *PDP1* was identified. Recombinant PDP1 protein was also used to generate a polyclonal antibody. This was used for Western blot analysis and subsequently showed a large reduction in PDP1 protein in patient cell lysates [3].

4.2 Development of PDP1 assay in canine tissue

Two years after the first genetic cause of PDP1 deficiency in humans was identified, Shelton sent tissue samples from four related Clumber Spaniels to the Robinson lab for investigation. The dogs suffered from profound exercise intolerance with elevations in lactate and pyruvate causing metabolic acidosis. Cameron and Robinson's lab coworkers performed PDH activity assays and restored PDH activity after incubation of cell lysates with recombinant PDP1 protein, as was performed for human tissue samples. Western blot analysis showed a complete loss of PDP1 expression. The affected dogs were found to be homozygous for a mutation in the *PDP1* gene. The mutation c.754C > T created a premature stop codon p.Q252X was identified by gene sequencing. The mutation in the *PDP1* gene destroyed a restriction site, and a genotypic test was created, all methodology as previously reported by us [4]. The test is known as RFLP or restriction fragment length polymorphism. The generation of this test allowed for the ability to genotype the Clumber spaniel for this mutation. A pedigree analysis for five generations is shown (**Figure 2**). A full pedigree of all dogs tested in the Robinson lab is shown for the first time (**Figure 3**). Affected dogs typically present between 15 weeks to 1 year old with exercise intolerance, described as the unwillingness to exercise or play after 5 minutes. Activity may resume after rest, but the dog will tire again quickly. Neonatal deaths have been reported in the littermates of affected dogs. Affected dogs are recommended to be fed a low-carbohydrate, high-fat diet. Carriers of the mutation do not show any clinical features [4].

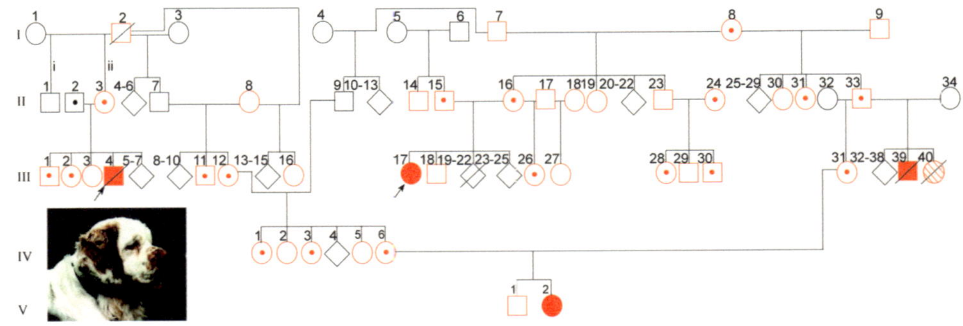

Figure 2.
Pedigree of five generations of clumber spaniels and genotyping for the c.754C > T PDP1 founder mutation. Red symbols represent individual dogs that were genotyped. Individuals of unknown sex are represented by diamonds. Carriers are identified with small circles within the symbol and homozygous affected individuals are represented by solid red symbols. The individual with red diagonal strips displayed clinical features but could not be tested. Separate breeding with the same parents is identified by small Roman numerals. This figure was originally published by us in molecular genetics and metabolism, vol 90, Cameron et al, identification of a canine model of pyruvate dehydrogenase phosphatase 1 deficiency, pages 15–23, copyright Elsevier, (2007) [4].

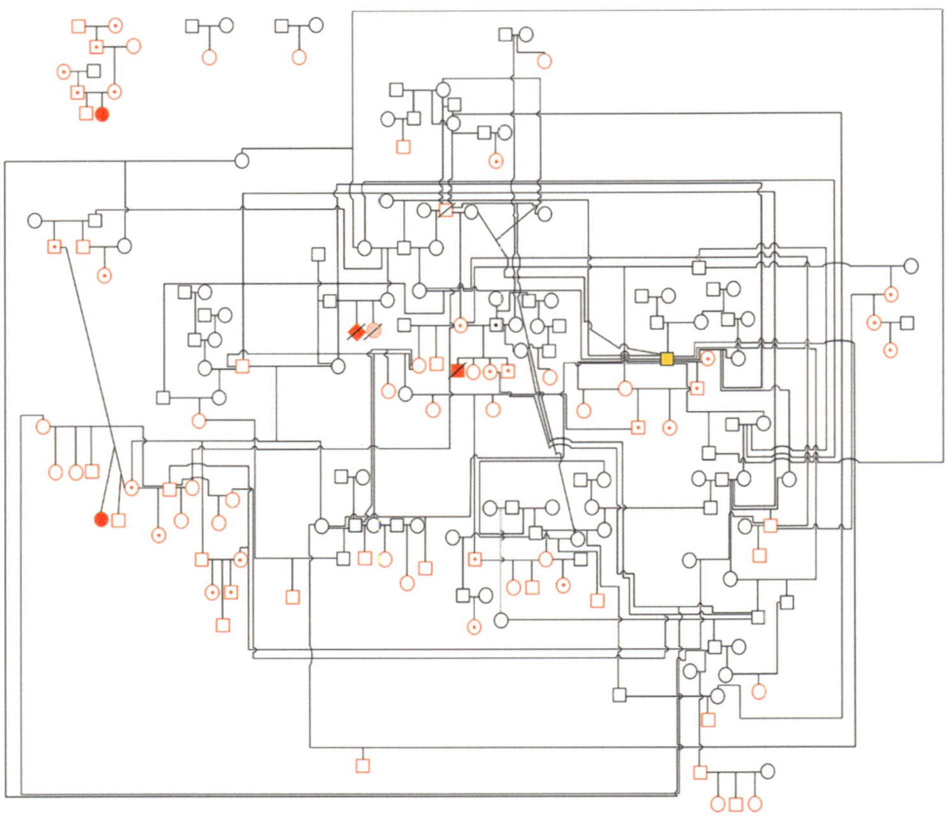

Figure 3.
Popular sire effect was seen in a small population of Clumber spaniels and genotyping for the c.754C > T PDP1 founder mutation. Red symbols represent individual dogs that were genotyped. Individuals of unknown sex are represented by diamonds. Carriers are identified with small circles within the symbol and homozygous affected individuals are represented by solid red symbols. The individual with red diagonal stripes displayed clinical features but could not be tested. The particular sire highlighted in gold is an example of a 'popular sire' and was mated 11 times within this pedigree. This data was collected during and after the development of the RFLP genetic test for the heritable PDP1 mutation in the Clumber spaniels [4]. All data have been de-identified to preserve the confidentiality of the animals and their owners.

5. Popular sire effect

The popular sire effect, also known as the popular sire syndrome, may occur when an animal with a certain trait or attribute becomes desirable. This sire will then be bred repeatedly with many females, causing a reduction of genetic diversity and allows for unconnected, undesirable genetic traits to spread rapidly. An example of selective breeding is akin to the Founder Effect that we attribute to humans. Dominant traits that have low reproductive fitness are typically removed from the gene. However, recessive traits are silently passed on until the two carriers of the recessive allele mate and have a homozygous recessive individual who is affected by a disorder. Carrier screening of the Clumber spaniel for the c.754C > T PDP1 founder mutation revealed an interesting pedigree shown below. The lines with multiple crossings highlight the complex level of consanguinity found in a subset of the breed (**Figure 3**). An example of a popular sire is highlighted by a gold square in **Figure 3**, where pedigree and genotyping analysis revealed that he was bred 11 times.

According to the Clumber Spaniel Club of America, the Clumber spaniel was one of the first breeds to be recognized by the American Kennel Club in 1884. In both the United States and the United Kingdom, it is estimated that approximately 200 Clumber puppies are registered per year. The dogs have an average lifespan of 10–12 years [23, 24]. Tests are easily available for PDP1 deficiency. The Canine Health Information Center of America reports that 362 PDP1 tests have been performed with 315 dogs being homozygous normal and 47 carriers identified [25]. Statistics from the 2018 Clumber spaniel breed health plan reports of 195 dogs tested, only 4 carriers have been identified. However, at least one parent of a breeding pair producing litters from 2015 to 2017 has not had their carrier status tested [23].

Screening of a second autosomal recessive disorder, referred to as Exercise Induced Collapse (EIC) began for Clumber spaniels in 2015. The mutation is in the dynamin 1 gene (*DNM1* gene) and was originally identified in Labrador retrievers by Patterson and coworkers in 2008, where linkage analysis of affected animals led to identifying the candidate gene. Sequencing of the *DNM1* gene of affected animals identified a mutation in a highly conserved region of the gene. The mutation c.767G > T alters a *SmlI* restriction site and a simple RFLP analysis can identify carriers and affected animals [26]. EIC is a neuromuscular disorder that presents as exercise intolerance in the absence of lactic acidosis. In contrast to the PDP1 disorder, EIC has a later onset of clinical features, at approximately 2 years of age and rarely results in early death, and does not appear to be completely penetrant. Clinical features include muscular weakness after 5–20 minutes of strenuous exercise, seen as the unsteady and uncoordinated gate of the hind limbs. Dogs typically recover from the episode within 15–30 minutes. In rare cases, EIC can progress to full-body weakness, confusion, loss of consciousness, and seizures [27]. Of the Clumbers 320 tested for EIC in the United Kingdom up to 2018, 40.8% were carriers and 4.7% were affected [23].

Pedigree analysis, linkage analysis, and modern techniques such as SNP array-based analysis of pure-breed dogs can be used to determine the coefficient of inbreeding (COI). COI refers to identical genetic information by descent and is calculated between 5 and 10 generations. It is the probability that two genes at any locus in an individual are derived from the identical diploid alleles of a previous generation [28]. For example, the COI for first cousins is 6.25%, half-siblings is 12.5%, and full siblings is 25%, provided the dam and sire do not pass through the same animal twice. For an autosomal loci, the inbreeding coefficient (F) of one individual is as shown [29]:

$$F = \left(\tfrac{1}{2}\right)^{(n_1 - n_2 + 1)} \tag{1}$$

where n_1 and n_2 are the numbers of generations that separate the individuals of the inbred population. An example of this calculation is shown in population genetics by Johnson, Keats, and Sherman, 2019, for first cousins related through two grandparents is $F = (1/2)^5 + (1/2)^5 = (1/2)^4 = 1/16$ or 6.25% [29].

In 2013, the COI for the Clumber spaniels in the United Kingdom was calculated to be 18.2%, which suggests low genetic diversity [30]. Repeated matings of the same dam to the same sire with complex consanguinity loops (**Figure 3**) are likely responsible for the loss of genetic diversity in this breed. In 2003, the UK Kennel Club recognized that the Clumber spaniel was a vulnerable British breed and was actively trying to decrease the COI by limiting the use of stud dogs and publishing the COI of breeding animals. A popular dam has a more limited effect on the COI, due to the constraints of the female reproductive cycle and litter count. With concerted effort and diligence, the COI for this breed had been reduced to 16.2% by 2017. The UK Kennel Club has been encouraging *PDP1* testing results to be included on its Mate Select pages [23].

In a small founder population, such as the Clumber spaniel, random genetic drift will increase relative to selection, where beneficial alleles can become lost and harmful alleles can increase in frequency in reference to larger populations. Current technologies allow investigators to study and identify harmful alleles in founder populations. In 2023, Donner and coworkers performed SNP array analysis for over one million dogs, with 242,665 purebred dogs and 811,628 mixed-breed dogs [31]. Their data, using an unparalleled population size and following over 250 disease-associated genetic variants, show that purebred dogs indeed have a lower mean heterozygosity level than mixed-breed dogs as mathematically predicted over 100 years ago by the pioneers of population genetics. This exhaustive study reinforces the need to maintain diversity by outcrossing programs to increase heterosis (hybrid vigor introduced by mating two unrelated purebred lines that each have the desirable trait) and reducing both the use of popular sires and repeated mating with the same dam.

The genotypes of the two known heritable autosomal recessive disorders affecting the Clumber spaniel can now be determined for all breeding animals. If all carriers are restricted from mating, one would assume that the disorders would be eradicated from the breed. However, many of the dogs being bred in both the United Kingdom and United States are not being genotyped and are referred to as "Heredity Clear" meaning that the disorders were not seen in the pedigree nor within the family history of the dog. In his 1967 edition of Introduction to Quantitative Genetics, Falconer calculates the number of generations required to effect a specific change in gene frequency in an unwanted recessive allele breeding program. For example, as calculated on page 36, if the incidence of a rare autosomal recessive disorder is 1/40,000, and simply restricting the mating of homozygous affected individuals, it would take 59 generations to reduce the frequency by half [32]. Therefore, if breeders of the Clumber spaniel fail to restrict breeding carriers, it is unlikely that the disease-causing alleles will be eradicated from the breed.

To address this concern, concerted efforts of kennel clubs around the world are encouraging the publication of carrier status for breeding Clumber spaniels, which is apparently reducing the reported incidence of both PDP1 deficiency and

EIC. However, due to the popular sire effect, the breed is approaching extinction if intervention is not taken. It concerns the effective population size (N_e) of the dogs. N_e is the number of breeding males and females in the population, according to the equation:

$$N_e = \frac{\left(4 \times N_m \times N_f\right)}{\left(N_m + N_f\right)} \quad (2)$$

where N_m is the number of breeding males and N_f is the number of breeding females [12]. In an effective population, N_e is 100 with a sex ratio of 1 (equal amount of breeding males to females). Due to the extreme usage of a popular sire, the Clumber spaniel has very restricted inbreeding, and the Effective Population Size (N_e) is currently measured at 24.5. When N_e is lower than 100, the population size is considered critical and below 50 suggests that a breed is close to extinction [23].

6. Conclusions

The concept of purebred dogs is thought to have been created in Victorian Britain, between 1837 and 1901. However, the origin of the Clumber spaniel can be traced back to the mid-1700s, when the 4th Duc de Noailles kept a unique kennel of dogs, he kept for hunting which he shared with the 2nd Duke of Newcastle. Historical documents describe how the French Revolution caused a bottleneck event for the breed, and the Clumbers we see today are ancestors from the kennel belonging to the Duke of Newcastle at Clumber Park and carefully selected for by his gamekeeper. The Clumber was rarely seen outside grand estates until 1859 when it became fashionable to own the breed as a show dog. In the 1900s, show breeders began to favor an increased weight for the breed, and between 1908 and 2008, the weight of the show dog doubled in size [2].

Radical selection to exaggerate any features is not in the best interests of any species, as first described in the theoretical works of pioneers in the field of Population Genetics, Bottleneck, and Founder Effects such as Darwin, Mendel, Garrod, and Mayr. Recent advances in genetics and genomics allow for the identification of genetic variations. Biochemical assays allow for the correlation between mutations and how they lead to enzyme deficiencies which can then be linked to the pathogenesis of disease. Once a disease-causing mutation is known, techniques can be developed that allow for genotyping of individuals and follow the heritability of the variant within a pedigree. These technologies now allow us to prove the mathematical theories of the past. With this knowledge comes reproductive choices. Luckily, the owners and breeders of the Clumber spaniel family of canines remain lovingly invested in preserving this wonderful and royal breed and have been choosing genetic testing to reduce the coefficient of inbreeding, and through genotype analysis to identify and discourage breeding of known carriers of *PDP1* and *DNM1* disorders. Giving context to well-established mathematical theories, through the odyssey of the Clumber spaniel, allows for a greater understanding of population genetics (**Figure 4** and **Table 1**).

Figure 4.
The above 7-month-old puppy was rescued from a puppy mill at 8 weeks old. He was the runt of the litter and has now been genotyped, testing positive for PDP1 deficiency. This picture is used with permission of Christy Edwards of North Carolina, USA, who is devoted to Clumber spaniel rescues.

Mating Type	Proportion of Shared Alleles	*F*
Parent-offspring	½	¼
Full siblings	½	¼
Half siblings	¼	1/8
Uncle-niece/Aunt-nephew	¼	1/8
First cousins	1/8	1/16
Double first cousins	¼	1/8
Half first cousins	1/16	1/32
First cousins once removed	1/16	1/32
Second cousins	1/32	1/64
Second cousins once removed	1/64	1/128
Third cousins	1/128	1/256

Table 1.
Proportion of alleles that are shared by related individuals, adapted from Population Genetics by Johnson et al. [29].

Acknowledgements

We would like to thank Christy Edwards of North Carolina, USA, rescuer of affected Clumber spaniels bred in puppy mills, for her valuable insights and unwavering devotion to this regal breed. Her organization is called Clumber Spaniel Rescues, a Private Breed Rescue.

We also thank James Darley of Buckinghamshire, England, for personal communications, the sharing of his detailed research and historical documentation regarding the Clumber, and his devotion to the rebirth of the Royal Spaniel.

Lastly, we thank Dr. Bryant Freeman, Kansas, USA, for personal communication and the sharing of his detailed historical documentation regarding the Clumber and his devotion to the Clumber Spaniel Club of America with the title of Breed Historian.

Christy, James, and Bryant gave much-appreciated depth to our understanding of the Clumber spaniel and the 250+ year odyssey the breed has taken, leading to the small part that our group played in identifying a genetic cause for disease which is leading to increasing the health of a most Royal and Loyal breed.

Conflict of interest

The authors declare no conflict of interest.

Author details

Felicia Ikolo[1†], Sabyasachi Maity[2†], Robert Finn[3†], Atoum Abdullah[1], Alireza Tajik[1], Jessie M. Cameron[4,5] and Mary C. Maj[1*]

1 Biochemistry Department, St. Georges School of Medicine, Grenada, West Indies

2 Physiology Department, St. Georges School of Medicine, Grenada, West Indies

3 Biochemistry Department, St. George's International Medical School, Newcastle Upon Tyne, UK

4 Department of Paediatric Laboratory Medicine, The Hospital for Sick Children, Toronto, Ontario, Canada

5 Department of Laboratory Medicine and Pathobiology, University of Toronto, Toronto, Ontario, Canada

*Address all correspondence to: mmaj@sgu.edu

†Authors contributed equally as first authors.

References

[1] Darwin C. The Origin of Species. Elliot CW, editor. New York: P.F. Collier and Son; 1909. p. 557

[2] Darley J. Rebirth of the Royal Spaniel: The Clumber in the Field. 1st ed. Oxford: Venaticus Collection; 2021

[3] Maj MC, MacKay N, Levandovskiy V, Addis J, Baumgartner ER, Baumgartner MR, et al. Pyruvate dehydrogenase phosphatase deficiency: Identification of the first mutation in two brothers and restoration of activity by protein complementation. The Journal of Clinical Endocrinology and Metabolism. 2005;**90**(7):4101-4107

[4] Cameron JM, Maj MC, Levandovskiy V, MacKay N, Shelton GD, Robinson BH. Identification of a canine model of pyruvate dehydrogenase phosphatase 1 deficiency. Molecular Genetics and Metabolism. 2007;**90**(1):15-23

[5] Mendel G. Versuche über Pflanzenhybriden. Verhandlungen des naturforschenden Vereines Brünn. 1866;**4**:3-47

[6] Wilkie J. Some reasons for the rediscovery and appreciation of Mendel's work in the first years of the present century. British Journal for the History of Science. 1962;**1**(1):5-17

[7] Garrod AE. The incidence of alkaptonuria: A study in chemical individuality. Lancet. 1902;**160**(4137): 1616-1620

[8] Bateson W. Mendel's Principles of Heredity: A Defence. Vol. 44. Cambridge University Press; 1902. pp. 1-39

[9] Weinberg W. Über den Nachweis der Vererbung beim Menschen. Naturkd Württ. 1908;**64**:369-382

[10] Hardy G. Mendilian proportions in a mixed population. Science. 1908;**28**(706):49-50

[11] Fisher R. The Genetical Theory of Natural Selection. Oxford: Clarendon; 1930

[12] Wright S. Evolution in mendelian populations. Genetics. 1931;**16**(2): 97-159

[13] Mayr E. Change of genetic environment and evolution. In: Huxley J, Hardy AC, Ford EB, editors. Evolution as a Process. London: George Allen and Unwin; 1954. pp. 157-180. Available from: http://www.sevin.ru/ fundecology/ speciation/mayr(20).pdf

[14] Ewans WJ. The probability of survival of a new mutant in a fluctuating environment. Heredity (Edinb) [Internet]. 1967;**22**:438-443. DOI: 10.1038/hdy.1967.53

[15] Crow JF, Kimura M. The effective number of a population with overlapping generations: A correction and further discussion. American Journal of Human Genetics. 1972;**24**(1):1-10

[16] Haudry A, Laurent S, Kapun M. Population genomics on the Fly: Recent advances in drosophila. Methods in Molecular Biology. 2020;**2090**:357-396

[17] Chase K, Jones P, Martin A, Ostrander EA, Lark KG. Genetic mapping of fixed phenotypes: Disease frequency as a breed characteristic. The Journal of Heredity. 2009;**100**(Suppl. 1):S37-S41

[18] Freeman B. Clumber Spaniels: Lord and Master of Clumber Park. AKC Gazette; 2023. pp. 49-52

[19] American Kennel Club [Internet]. Available from: https://www.akc.org/dog-breeds/clumber-spaniel/

[20] Herrtage M, Houlton I. Collapsing Clumber Spaniels. The Veterinary Record. 1979;**105**(14):334

[21] Järvinen AK, Sankari S. Lactic acidosis in a Clumber Spaniel. Acta Veterinaria Scandinavica. 1996;**37**(1):119-121

[22] Robinson BH, Sherwood WG. Pyruvate dehydrogenase phosphatase deficiency: A cause of congenital chronic lactic acidosis in infancy. Pediatric Research. 1975;**9**(12):935-939

[23] Clumber Spaniel Club. 2018 Clumber Spaniel Breed Health Plan [Internet]. 2018. p. 1-12. Available from: https://www.clumberspanielclub.org.uk/wp-content/uploads/2019/02/2018-Health-Plan.pdf

[24] Williams W. Clumber Spaniel: Dog Breed Characteristics & Care [Internet]. 2023. Available from: https://betterpet.com/clumber-spaniel/

[25] The Canine Health Information Center OFA Clumber Spaniel [Internet]. 2023. Available from: https://ofa.org/chic-programs/browse-by-breed/?breed=CLS [Accessed: Oct 4, 2023]

[26] Patterson E, Minor K, Tchernatynskaia A, Taylor S, Shelton G, Ekenstedt K, et al. A canine DNM1 mutation is highly associated with the syndrome of exercise-induced collapse. Nature Genetics [Internet]. 2008;**40**(10):1235-1239. Available from: https://pubmed.ncbi.nlm.nih.gov/18806795/

[27] Taylor SM, Shmon CL, Diane Shelton G, Patterson EE, Minor K, Mickelson JR. Exercise-induced collapse of Labrador retrievers: Survey results and preliminary investigation of heritability. Journal of the American Animal Hospital Association. 2008;**44**(6):295-301

[28] Wright S. Coefficients of inbreeding and relationship. The American Naturalist. 1922;**56**:330-338

[29] Johnson H, Keats J, Sherman S. Population genetics. In: Pyeritz R, Korf B, Grody W, editors. Emery and Rimoin's Principles and Practice of Medical Genetics and Genomics: Foundations. 7th ed. London, San Diego, Cambridge and Oxford: Elsevier; 2019. pp. 359-373

[30] Dog Breed Health: Clumber Spaniel [Internet]. 2023. Available from: https://www.dogbreedhealth.com/clumber-spaniel/

[31] Donner J, Freyer J, Davison S, Anderson H, Blades M, Honkanen L, et al. Genetic prevalence and clinical relevance of canine mendelian disease variants in over one million dogs. PLoS Genetics. [Internet]. 2023;**19**(2):1-26. DOI: 10.1371/journal.pgen.1010651

[32] Falconer DS. Introduction to Quantitative Genetics. 4th ed. Edinburg and London: Oliver and Boyd LTD; 1967. 345 p

Section 5

Oral Malignancies and Genetics

Chapter 8

Genetic Revelation of the Potentially Malignant Disorders in the Oral and Maxillofacial Region

Nitish Krishna Das, A.K.M. Shafiul Kadir, Mohammad Ullah Shemanto, Ety Akhter, Ashik Sharfaraz, Soumik Tripura, Joye Kundu and Ayesha Afrose Ura

Abstract

The oral and maxillofacial region is the site of numerous cancer forms. The most frequent cancer, which accounts for more than 90% of these tumors, is squamous cell carcinoma. Genetic changes caused by malignant transformation later result in phenotypic changes in cells. Potentially malignant disorders and circumstances can lead to the development of some malignancies, such as oral squamous cell carcinomas (OSCCs). Because OSCC and precursor lesions cannot be detected early, the 5-year survival rate for OSCC is still only about 50%. Early detection of oral cancer, particularly in the premalignant stage, can greatly reduce death and morbidity. The clinical, histological revelations and etiopathogenesis of a few potentially malignant disorders of the oral and maxillofacial region are reviewed in this chapter.

Keywords: mutation, genetic element, polymorphism, potentially malignant disorders, oral and maxillofacial region

1. Introduction

Cancer is a collection of disorders have been designated by unusual proliferation of cells and the capacity to migrate or spread to other regions of the body [1]. According to Global Cancer Statistics 2022, more than 19.3 million (19,300,000) new cancer cases were diagnosed and recorded recently, resulting in nearly 10 million fatalities by 2020 [2]. According to projections made by the American Cancer Society, it was estimated that in the year 2022, there would be a total of 1,918,030 newly diagnosed cases of cancer and 609,360 deaths attributed to cancer in the United States [3]. Oral and maxillofacial cancer is one of the most common kinds of cancer, and it is becoming a growing concern in numerous regions of the world [4]. Oral cancer is the sixteenth most prevalent malignancy and the fifteenth leading cause of death globally, with an incidence of four cases per 100,000 people (age-adjusted), with substantial regional differences depending on gender, age groups, nations, races and ethnicities, and socioeconomic conditions [5]. Geographical location is a significant role in the development of oral cancer since persons with poor socioeconomic class are more likely to get the disease because they are less aware of the dangers of avoidable risk factors such as tobacco, areca nut, and alcohol use [6]. In South Asian nations such as India, Sri Lanka, Pakistan, and Bangladesh, it is the most prevalent kind of oral cancer and accounts for around one-fourth of all new cases [4].

Oral cancer may originate through two primary pathways: *de novo* development or progression from potentially malignant disorders. These lesions commonly manifest as leukoplakia (characterized by a white patch), erythroplakia (characterized by a red patch), or erythroleukoplakia (characterized by a mixed red and white patch) [7]. Squamous cell carcinoma (OSCC) is the most prevalent cancer, accounting for more than 90% of all malignancies, and it develops from potentially malignant lesions and conditions in the oral and maxillofacial facial region [8]. There are several stages to the development of oral cancer from potentially malignant disorders, including genetic, epigenetic, and metabolic changes [5, 9–11]. Genetic alterations that arise during the process of carcinogenesis can manifest as various types of genomic changes, including point mutations, amplifications, rearrangements, and deletions [12]. Proto-oncogenes undergo a conversion process that leads to the formation of oncogenes. These oncogenes are responsible for modifying the protein sequences of growth-promoting proteins and factors that regulate the cell cycle. Consequently, uncontrolled cell division occurs weakening cell cohesion and causing local infiltration, which plays a crucial role in the progression from dysplasia to oral squamous cell carcinoma (OSCC) [13]. Numerous oncogenes play a vital role in cellular responses, and these include the proto-oncogene epidermal growth factor receptor (EGFR/c-erb 1), various members of the ras gene family, c-myc, int-2, hst-1, PRAD-1, bcl-1, and the growth regulator P53 [14].

The management of oral squamous cell carcinoma (OSCC) has undergone significant transformations; in recent years, primarily due to the emergence of immunotherapy, in addition to conventional approaches such as surgery, systemic therapy, and radiotherapy, either as individual therapies or in combination. Nevertheless, a significant proportion of cases are typically identified during the later stages of the ailment, resulting in a substantial reduction in the overall survival rate among affected individuals [15]. The most effective approach to preventing the transformation of cancer is through the early detection of potentially malignant disorders, which involves regular oral examinations and the removal of environmental and behavioral risk factors that are associated with such lesions [16, 17].

In this chapter, the epidemiology and clinico-histological aspects of numerous forms of potentially malignant disorders in the oral and maxillofacial region will be briefly discussed. Their risk factors, genetic and epigenetic underpinnings, early detection approaches, and therapy to prevent the development of malignancies will also be emphasized.

2. Methodology

The chapter was prepared by conducting a thorough review of the existing literature on potentially malignant disorders in the oral and maxillofacial region from the reputable scientific databases, including PubMed, Scopus, ScienceDirect, Google Scholar, and Cochrane using Boolean operators. The article search was carried out using relevant keywords. The inclusion criteria for preparing this chapter were articles about genetic aspects of OPMDs published online, articles in English, and articles with both experimental and observational research. Incomplete and inaccessible articles were excluded. Data retrieval began with selecting and determining literature that was relevant to the objectives and met the inclusion criteria. Next, literature that met the inclusion criteria was extracted. Data extracted from each piece of literature included author, year of publication, title, and journal name. Studies that specifically examined the genetic aspects of these disorders were carefully selected. Through a critical analysis of the chosen studies, the information was synthesized to provide a cohesive and informative overview of the genetic revelation of potentially malignant disorders in the oral and maxillofacial region. The methodology implemented here ensured the inclusion of proper references and allowed peers to reproduce the experiments and obtain consistent outcomes.

3. What are the potentially malignant disorders of the oral and maxillofacial region?

Usually, the clinical appearance regarding oral cancer is tremendously varied along with symptoms as well as signs, which are very often related to the primary tumor. Several cause are responsible to generate oral cancer for instance oral cavity lesion, ulcerated lesion, mobile teeth, bleeding, pain or numbness in the mouth or face, ill-fitting dental prosthesis could cause oral cancer. Leaving apart, viral lesions (warts), and even white lesions can be considered as malignant on biopsy. Erythematous (abnormal redness of mucous membrane) lesion, which is found during biopsy of erythroplakia, can also induce suspicion of oral cancer. However, various potentially malignant disorders of the oral and maxillofacial region are listed in the **Table 1** below [18]:

Some of the potentially malignant disorders of the oral and maxillofacial region, which has the potency for the malignant transformation will be discussed in this chapter.

4. Epidemiology of potentially malignant disorders of the oral and maxillofacial region

Understanding the epidemiology of potentially malignant disorders of the oral and maxillofacial region is crucial for effective prevention and management strategies.

Potentially malignant disorders of the oral region	Potentially malignant disorders of the maxillofacial region
Leukoplakia	Actinic cheilitis
Proliferative verrucous leukoplakia	Actinic keratosis
Erythroplakia	Epidermolysis bullosa
Oral submucous fibrosis	Dysplastic nevi
Oral lichen planus and oral lichenoid reaction	Congenital melanocytic nevus
Lichenoid dysplasia	Sebaceous Nevus

Table 1.
Potentially malignant disorders of the oral and maxillofacial region.

This summary explores the current updates on the epidemiology of potentially malignant disorders, including age, sex, changes in prevalence, geographical and demographic patterns, and the impact of public health interventions.

Age and Sex: Potentially malignant disorders can affect individuals of all age groups, with older adults being more susceptible due to cumulative exposure to risk factors, such as tobacco and alcohol use. Regular oral health screenings become increasingly important as individuals age to detect and manage lesions early. In terms of sex, studies suggest that males have a higher incidence of potentially malignant disorders, possibly due to higher rates of tobacco and alcohol consumption [19]. However, it is important to note that these lesions can affect individuals of any gender, emphasizing the need for preventive measures for everyone.

Changes in Prevalence: Over time, there have been changes in the prevalence of potentially malignant disorders. While tobacco and alcohol use remains a significant risk factor, some regions have seen a decrease in prevalence due to increased awareness, public health campaigns, and stricter regulations. However, these lesions still pose a significant burden globally, with increasing trends in certain populations [20]. This highlights the ongoing challenges in combating the impact of tobacco and alcohol on oral and maxillofacial health, necessitating continued preventive measures.

Geographical and Demographic Patterns: The epidemiology of potentially malignant disorders exhibits geographical and demographic variations. Regions, such as Southeast Asia and parts of Africa, have higher incidence rates due to specific risk factors prevalent in those areas, such as betel quid chewing. Certain demographic groups, such as older adults and those with low socioeconomic status, may have a higher prevalence due to disparities in healthcare access and lifestyle factors [21]. Understanding these patterns is crucial for implementing targeted prevention and early detection strategies.

Impact of Public Health Interventions: Public health interventions have played a significant role in reducing the burden of potentially malignant disorders. Tobacco control policies, including taxation, cessation programs, and awareness campaigns, have led to a decline in tobacco use and associated lesions in some populations. Similarly, alcohol control measures have shown positive effects [22]. Targeted efforts for high-risk populations, such as screening programs and education initiatives, have facilitated early detection and intervention [23]. By combining population-wide interventions with targeted strategies, further reductions in the incidence and prevalence of these lesions can be achieved [24].

It is crucial of understanding the epidemiology of potentially malignant disorders of the oral and maxillofacial region for effective prevention and management. Age and sex play roles in susceptibility, with older adults and males being more affected, although everyone is at risk. Changes in prevalence, geographical and demographic patterns, and the impact of public health interventions are important considerations. Continued research and collaborative efforts are necessary to address the dynamic nature of these lesions and reduce their global burden on public health.

5. Risk factors, which potentiate the malignant transformation of the potentially malignant disorders of the oral and maxillofacial region

Oral cancers are typically preceded by potentially malignant lesions [25, 26], and 50% precursor lesions converted to oral cancers [27]. Oral cancer exhibits significant variations in its incidence and prevalence across different regions worldwide,

Figure 1.
Risk factors for the malignant transformation of the potentially malignant disorders of the oral and maxillofacial region.

highlighting its distinct geographic distribution [28]. Oral cancer is a multifactorial lesion influenced by various factors. These risk factors encompass using tobacco, consuming alcohol, prolonged or excessive UV radiation exposure (generally labial carcinoma), viral infection (e.g., Human papillomavirus (HPV)), fungal infection (e.g., Candida), long-term inflammation, weakened immune system, genetic susceptibility, and dietary habits (**Figure 1**) [27].

The primary causes of oral cancer identified by various studies are using tobacco and consuming alcohol [26, 27, 29, 30]. The smoke from cigarettes contains various substances, including nicotine, which affects the central nervous system and the gastrointestinal tract, including the oral cavity [27]. Two nicotine metabolites, 4-(Methylnitrosamino)-1-(3-pyridyl)-1-butanone (NNK) and N′-nitrosonornicotine (NNN), possess carcinogenic properties [27]. These substances bind to the nicotinic acetylcholine receptor, promoting cell growth and creating a favorable environment for tumor development [27]. Smokeless tobacco (SLT) has also been significantly associated with potentially malignant and malignant oral cavity lesions [31]. The International Agency for Research on Cancer (IARC) monograph confirms a substantial link between SLT use and oral cancer. Marijuana smoke contains cannabinoids, immunosuppressants, and a mixture of potentially mutagenic compounds [32]. A case–control study on head and neck squamous cell carcinoma (HNSCC) found a negative correlation between marijuana use and oral cavity cancers [32]. Ethanol, when metabolized into acetaldehyde, a known carcinogen and tumor promoter, contributes to the development of oral cancer through chronic alcohol consumption [27]. Alcohol often contains carcinogenic impurities, such as polycyclic aromatic hydrocarbons and nitrosamines [27]. Although alcohol alone does not have a direct association with cancer progression, it synergistically interacts with tobacco to increase the risk of developing cancer [27]. According to the American Cancer Society, heavy smokers and drinkers face a risk of these cancers approximately 30 times higher than that of nonsmokers and nondrinkers. Betel quid chewing has been identified as a risk factor for oral precancerous lesions [33], particularly playing a major role in the development of submucosal fibrosis [30].

The oral microbiota plays a critical role in maintaining human health by contributing to immune response, nutrient digestion, and carcinogen metabolism [27]. It has been observed that patients with oral cancer often have poor oral hygiene [30]. The human oral cavity is colonized by more than 700 different bacterial species, collectively known as the oral microbiome [27]. Certain viruses such as HPV, EBV, Hepatitis B and C, Helicobacter pylori, as well as bacteria, such as *Porphyromonas*, *Fusobacterium nucleatum*, and *Prevotella intermedia,* have been identified as causes of various types of human cancer [27, 30, 34]. Recent studies found that some bacterial activity on oral epithelial cells, especially bacteria from periopathogenic biofilms, for example, *Fusobacterium nucleatum* and *Porphyromonas gingivalis* has the ability to create malignant conditions, which can influence oral oncogenesis [30]. Human papillomavirus (HPV) is a significant risk factor for a specific subset of head and neck squamous cell carcinomas (HNSCC) [27, 35–37]. Among HNSCC cases, HPV16 is the most prevalent type associated with carcinogenesis, followed by HPV18 [27]. HPV16 carries a higher risk for oropharyngeal squamous cell carcinoma (OPSCC). The early proteins of HPV, namely E6 and E7, play a crucial role in HPV-related OPSCC. E6 inhibits the tumor suppressor protein p53, while E7 binds to pRb (retinoblastoma protein) [27].

Oral cancer can develop as a result of chronic irritation in the mucosa [30]. The release of mediators (e.g., cytokines) is triggered by persistent inflammation, which is linked to chronic irritation and subsequently results in DNA damage due to oxidative stress, contributing in the development of cancer [30]. Chronic mucosal trauma is considered both an initiating factor and a progression promoter in oral cancer. It can either cause lesions on healthy mucosa or exacerbate existing lesions [30]. The risk of oral cancer is increased in individuals who undergo transplantation of allogeneic hematopoietic stem cells, particularly those who develop chronic graft-versus-host disease (GvHD) [38]. For the early detection and management of secondary malignant diseases, it is advisable to closely monitor patients who have chronic graft-versus-host disease (GvHD) on a regular basis. Bullous forms, plaque-like, ulcerative, erythematous erosions, papular, and white reticular striations are the subtype of oral lichen planus (OLP), a chronic inflammatory disease [39]. Those subtypes can affect any person singularly as well combined. OLP promotes some major cancer hallmarks such as immune evasion and tumor-promoting inflammation, and significantly disturbs the immune system [39].

A genetic predisposition plays a significant role in the development of oral squamous cell carcinoma (OSCC), particularly in tongue and buccal mucosa cancers [27]. However, determining the specific genetic or familial disposition for oral cancer is challenging due to the presence of concurrent risk factors, such as smoking and alcohol use [27]. Aberrant methylation has been observed as an early molecular event in the process of oral carcinogenesis [29]. For instance, gene-specific studies have identified inactivation of the p16INK4a gene through CpG methylation as a crucial event in epithelial dysplasia. Additionally, p16INK4a hypermethylation has been associated with loss of heterozygosity (LOH) [29]. Although genetic loss of heterozygosity is considered a better marker for predicting the risk of malignant progression in potentially malignant oral epithelial lesions (PPOEL), it has not yet been integrated into routine clinical practice [25]. Approximately 86 genes, along with the long interspersed elements 1 (LINE1), have been identified to exhibit hypermethylation associated with oral cancer [29]. Most of these 86 hypermethylated genes have been found to be downregulated and exhibit hypermethylation in their promoter regions in OSCCs compared to normal tissue [29]. Some researchers propose that individuals who inherit an impaired ability to metabolize carcinogens or procarcinogens and repair DNA damage are more susceptible to developing oral malignancies [27].

The risk of developing oral cancer may be increased due to a low fruit and vegetable diet, coupled with a deficiency in vitamin A. Dietary habits have been consistently linked to the development of several types of cancers [40]. Numerous studies seem to indicate that different food compounds could alter or modify cancer cells [40]. In 2019, a systematic study report published by the World Cancer Research Fund (WCRF) suggested that cancer, for example, oral cancer can be reduced by taking vegetables, fruit, whole grains, and legumes rich diet [41]. Submucosal fibrosis, an oral potentially malignant lesion (OPL), was initially considered idiopathic but is now believed to have a multifactorial etiology. Factors, such as capsaicin, present in chilies, as well as deficiencies in iron, zinc, and vitamins, have been implicated [30]. Consuming very hot foods has been identified as a risk factor for potentially malignant disorders, as a study found a significant difference between patients with oral lichen planus (OLP) and a control group [42]. Oxidative stress and antioxidants are important indicators in oral cancer and potentially malignant disorders as they may predict susceptibility to the development of oral cancer. Normal individuals exhibit differences in the levels of both enzymatic and nonenzymatic antioxidants when compared to patients with oral cancer or oral potentially malignant lesions [43]. Additionally, patients with oral potentially malignant lesions or oral cancer show high level of lipid peroxidation and higher oxidative stress than healthy people [43].

Men have a higher tendency to develop oral cancer compared to women. Out of every 100,000 men, approximately 5.8 men are at risk of being affected by oral cancer, whereas the corresponding number for women is 2.3. The rates of tobacco use and alcohol consumption have recently increased among women, which is significantly changing the current rate of women oral cancer [28]. Due to the difference in lifestyle of men and women, the influence of other risk factors on oral cancer and potentially malignant lesions remains unexplained [44]. There may be some undiscovered factors. There is a fluctuation in women's oral cancer rate and tumor occurring site, although men have consistency, which lifestyle alone can not explain [44]. Diabetic and underweight individuals were found to be at increased risk of oral potentially malignant lesions [26].

6. Clinico-histological features of the potentially malignant disorders of the oral and maxillofacial region

The clinical presentations and histological features of some of the potentially malignant disorders of the oral and maxillofacial region, which has the potency for the malignant transformation will be discussed in this part. The clinico-histological features are summarized in **Table 2**.

Potentially malignant disorders of the oral region	Affected area	Clinical presentations	Histological features
Leukoplakia	Any mucous membranes of the oral region.	White or grayish patches in the oral mucosa, well-defined or irregular border. Carcinoma *in situ* can be found	Hyperkeratosis and epithelial dysplasia can be seen.
Erythroplakia	Buccal and palatal mucous membrane of the oral region, including lower surface of the tongue and floor of the mouth.	Red patches with smooth texture in the oral mucosa, ill-defined border.	Epithelial hyperplasia with cellular atypia and nuclear polymorphism is present. Vascular dilatation, epithelial dysjunction, and basement membrane alteration can be seen.
Oral submucous fibrosis (OSF)	Any mucous membranes of the oral region.	Reduced mouth opening due to progressive mucosal fibrosis, inability to consume hot and spicy food due to burning sensation in the oral mucosa, reduced flexibility of the buccal and labial mucosa	Thinning of epithelial layers, epithelial dysplasia, subepithelial connective tissue fibrosis, hyalinization of the blood vessels, and inflammatory infiltrate may also be seen.
Oral lichen planus (OLP)	Buccal, gingival, and tongue mucosa of the oral region	Wickham's striae are seen in reticular form. Dysgeusia, burning sensation in the mouth is the common presentations.	Acanthosis (epithelial thickening), hyperkeratosis (epithelial thickening on the outside), and uneven or sawtooth rete ridges (elongation and thickening of the ridges between epithelial rete pegs) are present. Hypergranulosis, subepithelial fibrosis, basal cell degeneration, inflammatory infiltrate, and civatte bodies can be seen.

Table 2.
Clinico-histological features of OPMDs.

6.1 Leukoplakia

6.1.1 Clinical presentations

White or Grayish Patches: The primary characteristic of leukoplakia is the presence of white or grayish patches or plaques on the oral mucosa. These patches can vary in size, shape, and texture [18].

Non-Removability: Leukoplakia patches cannot be scraped off or wiped away. They are firmly attached to the underlying tissues and persist despite gentle manipulation [45].

Irregular Borders: The borders of leukoplakia patches may appear well-defined or irregular [46].

Surface Texture: The surface of the patches can be smooth, rough, or fissured [46].

Thickness: Leukoplakia patches may be flat or slightly raised and can have a thickened consistency [46].

Asymptomatic: In many cases, leukoplakia does not cause any symptoms and is discovered during routine oral examinations. However, some individuals may experience a burning sensation or sensitivity to spicy foods [47].

6.1.2 Histopathological findings

Histopathological examination of leukoplakia is essential for confirming the diagnosis and assessing the degree of dysplasia (abnormal cell changes). The histopathological features of leukoplakia can vary depending on the severity of dysplasia and the specific characteristics of the lesion. Here are the detailed histopathological features commonly observed in leukoplakia [45–47]:

Hyperkeratosis: Hyperkeratosis refers to an excessive accumulation of keratin, the tough, protective protein found in the outermost layer of the skin. In leukoplakia, there is often hyperkeratosis present in the epithelial layer of the lesion. This thickening of the superficial layer contributes to the characteristic white appearance of leukoplakia.

Acanthosis: Acanthosis refers to the thickening of the epithelial layer due to an increase in the number of cell layers. In leukoplakia, acanthosis can be observed in varying degrees. It indicates cellular proliferation and an alteration in the normal architecture of the oral mucosa.

Epithelial Dysplasia: Dysplasia refers to abnormal changes in the size, shape, and organization of cells within the epithelium. In leukoplakia, varying degrees of dysplasia can be observed, ranging from mild to severe. The presence of dysplasia is an important indicator of the potential for malignant transformation.

a. Mild Dysplasia: Mild dysplasia shows slight cellular abnormalities, including enlarged and hyperchromatic nuclei (increased nuclear size and increased staining intensity). The changes are limited to the lower third of the epithelium.

b. Moderate Dysplasia: Moderate dysplasia exhibits more pronounced cellular changes, extending beyond the lower third of the epithelium but not involving the full thickness.

c. Severe Dysplasia/Carcinoma in situ: Severe dysplasia, also known as carcinoma *in situ*, demonstrates marked cellular atypia, loss of normal cell differentiation, and involvement of the full thickness of the epithelium. However, there is no invasion into the underlying connective tissue.

Inflammatory Infiltrate: Inflammatory cells, such as lymphocytes and plasma cells, may be present within the underlying connective tissue of leukoplakia. The presence and extent of inflammation can vary and may indicate the body's response to dysplastic changes or other underlying factors.

It is important to note that the histopathological features of leukoplakia can vary from case to case. Additionally, the degree of dysplasia observed in the histopathological examination plays a significant role in determining the potential for malignant transformation. Regular histopathological evaluation and close clinical monitoring are necessary for appropriate management and treatment decisions in individuals with leukoplakia.

6.2 Erythroplakia

6.2.1 Clinical presentations

Erythroplakia is a potentially malignant oral lesion characterized by a distinct red patch or plaque on the mucous membranes of the oral cavity. It is present as a clinical entity that requires thorough evaluation and monitoring due to its high risk of malignant transformation. Here is a detailed description of the clinical presentation of erythroplakia [45–47]:

Erythematous Appearance: Erythroplakia appears as a persistent, bright red or deep red patch, or plaque on the oral mucosa. The red coloration is due to various factors, such as increased blood vessel dilation, inflammation, and changes in the epithelial cells.

Smooth or Granular Texture: The surface of erythroplakia can range from smooth and velvety to granular or irregular. It may have a slightly raised or flat appearance.

Non-Removable: Unlike some benign conditions, the red patch of erythroplakia cannot be easily wiped off or scraped away. It is firmly attached to the underlying tissues.

Borders: The borders of erythroplakia lesions are typically irregular or ill-defined. They may blend with the surrounding normal tissue or have a distinct demarcation. The border irregularity is an important characteristic that distinguishes erythroplakia from normal or benign oral mucosa.

Size and Shape: Erythroplakia can range in size from a few millimeters to several centimeters in diameter. The size of the lesion is not indicative of its malignant potential. Erythroplakia can have various shapes, including irregular, oval, or irregularly outlined.

Sites of Involvement: Erythroplakia commonly affects the dorsal surface of the tongue, particularly the posterior and lateral borders. It may also involve the ventral surface or appear as a patch underneath the tongue. The area beneath the tongue, known as the floor of the mouth, is another common site for erythroplakia. Erythroplakia may rarely involve the soft palate at the back of the mouth or the tonsillar area.

Asymptomatic: Erythroplakia is often asymptomatic in its early stages. It may be discovered incidentally during routine dental examinations or self-examinations. As the lesion progresses or becomes more advanced, individuals may experience slight discomfort, including a burning sensation or tenderness at the site of erythroplakia. In some cases, erythroplakia may bleed spontaneously or after minor trauma. However, bleeding is not a consistent feature and may occur in advanced or ulcerated lesions.

6.2.2 Histopathological findings

Histopathological examination of erythroplakia is crucial for assessing its malignant potential and guiding appropriate management. The histopathological characteristics of erythroplakia reveal cellular and tissue changes that indicate its status and progression toward malignancy. Here is a detailed description of the histopathological features commonly observed in erythroplakia [18, 45–47]:

Epithelial Hyperplasia: Erythroplakia often demonstrates varying degrees of epithelial hyperplasia, characterized by an increase in the number of epithelial layers. The basal layer may show increased cellularity and abnormal cell proliferation. The maturation process of the epithelial cells may be disturbed, leading to a lack of normal differentiation from basal to superficial layers. The epithelium may exhibit a disordered or irregular arrangement of cells with loss of normal cell maturation.

Dysplasia: Dysplastic changes are commonly observed in erythroplakia and indicate a higher risk of malignant transformation. Dysplasia refers to the presence of abnormal cellular and nuclear features within the epithelial layers.

Cellular Atypia: Dysplastic cells exhibit varying degrees of cellular atypia, including increased nuclear-to-cytoplasmic ratio, nuclear pleomorphism (variation in size and shape), hyperchromatism (increased nuclear staining), and increased mitotic activity.

Loss of Architectural Orientation: The dysplastic epithelium may demonstrate loss of normal architectural orientation, with a disorderly arrangement of cells and loss of polarity.

Chronic Inflammation: Inflammatory infiltrates, mainly consisting of lymphocytes and plasma cells, may be present in the connective tissue beneath the dysplastic epithelium. Chronic inflammation is a common feature due to the ongoing tissue reaction to the potentially malignant process.

Vascular Dilatation: Erythroplakia is characterized by a red appearance, indicating increased vascularity. Histopathologically, this is manifested by dilated and congested blood vessels within the lamina propria of epithelial-connective tissue interface.

Epithelial Dysjunction: The junction between the dysplastic epithelium and the underlying connective tissue may show irregularities or epithelial dysjunction. This loss of adhesion between the epithelium and connective tissue is indicative of potential invasive behavior.

Basement Membrane Alterations: Erythroplakia may demonstrate alterations in the basement membrane, which serves as a barrier between the epithelium and connective tissue. These alterations may include thickening, fragmentation, or disruptions in the basement membrane structure.

In advanced cases, erythroplakia may exhibit features of carcinoma *in situ*, where dysplastic changes involve the full thickness of the epithelium without invasion into the underlying connective tissue. Histopathological examination of erythroplakia helps determine the severity of dysplasia, identify any signs of invasive behavior, and guide appropriate management decisions, such as close monitoring, surgical excision, or other interventions.

6.3 Oral submucous fibrosis (OSF)

6.3.1 Clinical presentations

Clinical manifestations of oral submucous fibrosis (OSF) involve various changes in the oral cavity, including the mucosa, underlying connective tissue, and associated structures. The severity and progression of these manifestations can vary among individuals. Here are the detailed clinical manifestations and oral involvement commonly observed in OSF [18, 45–47]:

Mucosal Changes: Trismus is one of the hallmark features of OSF. As the fibrosis progresses, the fibrous bands in the oral mucosa cause a gradual reduction in mouth opening. Severe cases may result in significant limitation of jaw movement, making it difficult to eat, speak, or perform oral hygiene procedures. The oral mucosa affected by OSF often appears pale or opaque due to the fibrotic changes and reduced vascularity. The mucosa may have a leathery texture upon palpation. The fibrotic bands in OSF lead to a loss of elasticity in the oral mucosa. The affected tissues feel rigid and less pliable compared to normal mucosa.

Gingival Changes: The gingival tissues may become fibrotic and exhibit a firm leathery consistency. This fibrosis can lead to gingival recession, increased attachment loss, and impaired periodontal health. Fibrotic gingiva may restrict proper tooth brushing and flossing, leading to difficulties in maintaining oral hygiene. This can contribute to the progression of oral diseases, such as dental caries and periodontal disease.

Tongue Involvement: The tongue can be affected by fibrosis in OSF, resulting in reduced tongue mobility and firm consistency. This can lead to difficulties in speaking, swallowing, and articulating certain sounds. The filiform papillae on the dorsum of the tongue may progressively disappear due to fibrotic changes. This can result in a smooth or depapillated appearance of the tongue surface.

Cheek and Lip Changes: The fibrotic changes in OSF can affect the buccal mucosa, leading to a firm, rigid consistency, and reduced flexibility. This can result in difficulty in mouth opening, speaking, and chewing. Fibrosis of the lip mucosa may occur in OSF, causing a tightening or puckering of the lip tissues. This can lead to difficulty in lip movement, affecting functions such as smiling and speaking.

Other Manifestations: Many individuals with OSF experience a burning sensation or discomfort in the affected areas, particularly while consuming spicy or hot foods. In advanced stages of OSF, the fibrotic oral mucosa may become fragile and prone to ulceration. These ulcers can cause pain and increase the risk of secondary infections.

Malignant Transformation: OSF carries a potentially malignant nature, with an increased risk of oral cancer development. Oral leukoplakia, erythroplakia, and other dysplastic changes may be observed in OSF, indicating a higher risk of malignant transformation.

6.3.2 Histopathological findings

Histopathological examination of tissues affected by oral submucous fibrosis (OSF) reveals characteristic findings that help in confirming the diagnosis. Here are the detailed histopathological findings commonly observed in OSF [18, 45–47]:

Epithelial Changes: The epithelium may show varying degrees of atrophy, characterized by thinning of the epithelial layers. The superficial layers of the epithelium often exhibit hyperkeratosis, with an increased accumulation of keratinized cells. Epithelial dysplasia, characterized by cellular and nuclear atypia, may be observed in some cases of OSF. Dysplasia is considered a premalignant condition and indicates an increased risk of malignant transformation.

Fibrosis and Connective Tissue Changes: One of the key histopathological features of OSF is the presence of dense fibrosis in the subepithelial connective tissue. Fibrosis is mainly composed of collagen fibers and may extend into the deeper connective tissue layers. The collagen fibers in the fibrotic areas appear hyalinized and densely packed. They replace the normal loose connective tissue architecture. Increased numbers of fibroblasts, along with myofibroblasts, can be observed within the fibrotic areas. These cells are responsible for the excessive production and deposition of collagen. Vascular changes, such as hyalinization and obliteration of blood vessels, may be seen in the fibrotic regions. These changes contribute to reduced vascularity and impaired tissue nutrition.

Inflammatory Infiltrate: A chronic inflammatory infiltration composed of lymphocytes, plasma cells, and occasionally, eosinophils, is often present in the subepithelial region. This inflammatory infiltrate is believed to contribute to the fibrogenic process in OSF. Various pro-inflammatory cytokines and growth factors, such as transforming growth factor-beta, interleukin-6, and tumor necrosis factor-alpha, are upregulated in OSF. These molecules play a role in fibroblast activation and collagen synthesis.

Basement Membrane Changes: The basement membrane underlying the epithelium may show thickening and hyalinization. This alteration is believed to be a result of collagen deposition and fibrotic changes.

It is important to note that the severity and extent of these histopathological findings can vary among different individuals and at different stages of OSF. The presence of subepithelial fibrosis, along with epithelial changes and chronic inflammation, is considered essential for the diagnosis of OSF. However, histopathological examination alone may not be sufficient, and a correlation with clinical findings is necessary for accurate diagnosis and management.

6.4 Oral lichen planus (OLP)

Oral Lichen Planus (OLP) is a persistent inflammation of the buccal, lingual, palatal, and gingival mucosa of the oral cavity. It occurs when one's natural immunity erroneously assaults the healthy epithelium of the oral cavity, causing inflammation and tissue damage. OLP is a kind of lichen planus that can affect the skin, nails, scalp, and genitalia [48, 49].

6.4.1 Clinical presentations

Oral lichen planus (OLP) can manifest in a variety of clinical manifestations that differ from person to person. However, OLP has the following clinical representations [50]:

Reticular Pattern: The reticular form of OLP is the most common, and it is distinguished by a white, lacy, or web-like pattern on the oral mucosa. Wickham's striae are patterns that can form on the inner lining of the cheeks, tongue, gums, and other regions. The reticular pattern is usually asymptomatic and can be discovered by chance during a standard dental examination.

Erosive/Ulcerative Lesions: OLP can cause erosions or ulcers on the oral mucosa in some circumstances. These areas may appear red, raw, or eroded and may cause pain, burning sensations, or discomfort, especially when eating spicy or acidic meals. The erosive form of OLP might make it difficult to speak or swallow.

Atrophic Lesions: Atrophic OLP is distinguished by oral mucosa thinning and smoothness. The affected areas may appear glossy or shiny, as well as white or erythematous. Atrophic lesions are frequently related with food or drink sensitivities.

Bullous Lesions: In rare situations, OLP may show blister-like lesions known as bullae, which are fluid-filled sacs that can burst and leave erosions or ulcers behind. Bullous OLP can be unpleasant and cause discomfort or problems with oral function.

Plaque-like Lesions: Plaque-like lesions, which are raised areas of the oral mucosa, may appear in some people with OLP. These lesions may be white or gray in appearance and may cause discomfort or irritation.

Oral Symptoms: OLP can cause a burning or stinging sensation, dryness or changed salivary flow, altered taste sensation (dysgeusia), or the sensation of a foreign body in the mouth.

Gingival Involvement: OLP can damage the gingiva (gums) in some situations, causing redness, irritation, and desquamation (peeling) of the gum tissue.

6.4.2 Histopathological findings

The histopathological investigation is critical in diagnosing oral lichen planus (OLP) and distinguishing it from other oral diseases. The primary histopathological findings linked with OLP are as follows [18, 45, 50]:

Epithelial Changes: OLP is characterized by different epithelial alterations in the oral mucosa. These changes may include acanthosis (epithelial thickening), hyperkeratosis (epithelial thickening on the outside), and uneven or sawtooth rete ridges (elongation and thickening of the ridges between epithelial rete pegs).

Basal Cell Degeneration: The presence of basal cell degeneration is one of the defining histological markers of OLP. This is caused by liquefaction degeneration of the basal cell layer, which is characterized by basal cell adhesion loss and disruption of the basal cell layer's normal architecture.

Inflammatory Infiltrate: OLP is distinguished by a thick and band-like inflammatory infiltrate in the oral mucosa's underlying connective tissue. T-lymphocytes make up the majority of the infiltration, along with other immune cells such as macrophages and the occasional plasma cell. The inflammatory infiltration is frequently found in the subepithelial region and extends into the connective tissue papillae.

Civatte Bodies: Civatte bodies, also known as colloid bodies or apoptotic bodies, are histological abnormalities that are common in OLP. They are eosinophilic, tiny, spherical formations that reflect degenerated or apoptotic epithelial cells. These entities can be seen distributed throughout the epithelium and the underlying connective tissue.

Hypergranulosis: Hypergranulosis is the thickening and prominence of the granular layer in the epithelium. This feature is seen in OLP and is linked to aberrant keratinization of the oral mucosa.

Subepithelial Fibrosis: Subepithelial fibrosis may be found in more chronic or long-term cases of OLP. This is caused by the deposition of collagen fibers behind the epithelium, resulting in a thicker and fibrotic look.

It is vital to note that the histopathological findings in OLP differ depending on the disease's stage and severity. Furthermore, because some of these characteristics can be seen in other oral illnesses, histological testing is an important tool for correct diagnosis and differentiation from other disorders.

OLP is often diagnosed based on clinical characteristics along with histological results. For an appropriate diagnosis, a qualified pathologist with experience in oral pathology must assess the biopsy material and link the histological findings with the clinical presentation.

Understanding genetic alterations in the malignant transformation of the potentially malignant disorders of the oral and maxillofacial region.

The transformation of potentially malignant disorders into malignancies is heavily influenced by genetic alterations. These disorders, alternatively termed precursor lesions or intraepithelial neoplasia, encompass abnormal changes in tissue that possess the capacity to progress into cancerous conditions if not appropriately addressed and treated.

Oral cancer typically develops due to cellular alterations. When multiple epithelial cells undergo genetic changes, it is referred to as field cancerization. The initiation of this process is often triggered by oral lesions, which can be categorized as oral potentially malignant disorders. These potentially malignant lesions exhibit genetic instability, resulting in gains or losses of chromosomal material and alterations in nucleotide sequences. Even the border of malignant lesions, some oral cancer specimens show genetic changes. Over the past decade, research has highlighted the crucial role of aberrant DNA methylation in the development of oral cancer. However, oral carcinoma is influenced by various factors related to genetics [51].

During the progression from a potentially malignant lesion to cancer, several genetic alterations can accumulate, leading to uncontrolled cell growth, impaired cell death, and other hallmarks of malignancy. Here are essential insights concerning genetic alterations in the process of turning potentially malignant disorders into malignancies:

1. Oncogenes: Oncogenes are modified versions of normal genes, known as proto-oncogenes, responsible for regulating cell growth and division. Genetic changes, such as mutations, amplifications, or translocations, can transform proto-oncogenes into active oncogenes, leading to continuous activation. This persistent activation drives increased cell proliferation and survival, facilitating the progression of potentially malignant disorders into cancer.

   Oncogenes can be classified in accordance with the functions of their normal association (proto-oncogenes) in the biochemical pathways that controls sequestration and growth. These include in the following table (**Table 3**) [13]:

Oncogene category	Functions	Examples	Dysregulation/Mutational impact
Cell surface receptors	Regulate cell signaling and growth	EGFR, FGFR	Overexpression or mutations can lead to uncontrolled cell growth and cancer progression.
Growth factors	Stimulate cell proliferation	TGF, FGF, PDGF	Excessive production can promote cell proliferation, contributing to tumor development.
Inhibitors of apoptosis	Prevent cell death	BCL2	Overexpression can inhibit apoptosis, allowing abnormal cells to survive and grow.
DNA binding nuclear proteins	Regulate gene expression and transcription	MYC, FOS, JUN	Mutations can lead to increased cell proliferation and decreased apoptosis.
Cell cycle proteins	Control cell cycle progression	Cyclins and cyclin-dependent protein kinases	Dysregulation can disrupt cell cycle control, leading to uncontrolled cell division.
Intracellular signal transduction	Transmit signals within the cell	RAS	Mutations can lead to continuous activation of growth-promoting pathways.

Table 3.
Classification of oncogenes with their function and mutational impact.

2. Growth Factor Receptors: In human tumors, the activation of growth factor receptors involves gene rearrangements, mutations, and overexpression. The development of cancer and stem cells is influenced by specific signaling pathways such as the MAP-Kinase/ERK pathway, the JAK/STAT pathway, the NOTCH signaling pathway, the PI3K/AKT pathway, the Wnt pathway, and the TGF-β pathways.

   Dysregulation of growth factor receptors in oral carcinogenesis occurs due to increased production and autocrine stimulation. Carcinogenesis is characterized by irregular expression of transforming growth factor alpha (TGF-α) and beta (TGF-β) [13].

3. Tumor Suppressor Genes: The transformation of a potentially malignant cell into a malignant one occurs when tumor suppressor genes are inactivated, which is a significant event driving the development of malignancy [13]. These genes play a vital role in regulating cell growth and preventing cancer formation. In potentially malignant disorders, genetic changes can lead to the inactivation or deletion of tumor suppressor genes. Loss-of-function mutations in these genes can disrupt the control mechanisms that inhibit cell division, raising the probability of malignant transformation [39].

4. DNA Repair Genes: DNA repair genes are vital for maintaining genomic stability by repairing DNA damage. Genetic alterations affecting these genes can lead to genomic instability, contributing to the malignant transformation of potentially malignant disorders [52].

5. Chromosomal Abnormalities: Potentially malignant disorders have the potential to experience chromosomal alterations, including deletions, duplications, inversions, or translocations. These changes in the chromosomal structure can lead to disruptions in normal gene function and alter the patterns of gene expression, thereby playing a role in the advancement of cancer [53].

6. Mitochondrial Alteration: Mutations in the mitochondrial genome occur more rapidly than in the nuclear genome in response to oxidative damage and stress. It is speculated that an elevation in the number of copies of mitochondrial DNA serves as a cellular response to counteract dysfunction resulting from oxidative DNA damage. Studies have revealed an augmented copy number of mitochondrial genes in vitiligo, correlating with the severity of histopathological changes [54].

7. Epigenetic Modifications: Epigenetic modifications involve changes in gene expression without altering the DNA sequence. Aberrant epigenetic changes, such as DNA methylation or histone modifications, can silence tumor suppressor genes or activate oncogenes in potentially malignant disorders, thereby facilitating the transformation into malignancy. Epigenetic alterations encompass heritable changes in gene expression that do not involve modifications to the DNA sequence. These changes can occur more frequently than gene mutations and may persist not only throughout the cell's life but also across subsequent generations [22].

Gene silencing occurs when high-level expression changes due to altered gene transcription, driven by the influence of epimutations. This alteration disrupts the interplay between activators and suppressors on specific promoters within the context of chromatin. Consequently, disruptions in epigenetic mechanisms lead to the development of carcinomas and other epigenetic disorders by causing inappropriate gene expression [51].

8. DNA Methylation in the Development of Oral Cancer: DNA methylation is a cellular procedure, where the methyl groups are attached with the DNA molecule. Usually, methylation has the ability to alter the activity of a DNA segment except modifying the sequence. Thereafter, DNA methylation can not only influence inappropriate silencing of tumor suppressors but also associated to express oncogenes, which further induce cancer and other disorder. The generation of tumor cells is produced due to chromosome vulnerability through increased mutation rates, which is caused by the repression of proto-oncogenes transcription by either hypomethylation or demethylation. Hypermethylation also expresses the genetic event called loss of heterozygosity that promotes tumor growth. It is also associated with metastatic events and tumor neoangiogenesis [54].

9. DNA Aneuploidy in Cancer: In the context of cancer, genetically stable diploid cells undergo a transformation, giving rise to unstable aneuploid cells. In oral squamous cell carcinoma, the presence of DNA aneuploidy has been extensively studied using advanced techniques, such as flow cytometry and imaging. The findings from these studies reveal the existence of aneuploid tumor populations in numerous cases, underscoring the significance of ploidy status as a crucial prognostic factor [54].

10. Loss of Heterozygosity (LOH): The loss of genomic material from one of the chromosome pairs is called loss of heterozygosity (LOH). LOH in chromosomal regions thought to contain tumor suppressor genes may be associated with malignant developmental processes. LOH in oral potentially malignant disorders and its possible predictive value were recently reviewed. LOHs, particularly those located on chromosome arms 3p and 9p, have been shown to have a greater potential for malignant development of potentially malignant lesions [55].

11. p53: p53 involves many fundamental cell processes such as cell cycle progression, cellular differentiation, DNA repair, and apoptosis. The normal p53 protein has a very short half-life; the quantity in normal cells is extremely small. Mutant p53 protein has a long half-life and can accumulate to detectable levels in cells. This mutant protein is normally inactive, resulting in the loss of the protein's tumor suppressor function. More than 50% of oral squamous cell carcinomas are p53 positive, and mutations in the p53 gene have been documented [54]. When P53 is mutated, any lesions transform into oral cancers. Leaving apart, the loss of p53 function alters the ability of cells to respond to stress or damage and further promotes genomic instability, and thus promotes malignancy [52].

12. RASSF1 and RASSF2: Both RASSF1 and RASSF2 are members of the Ras family (RASSF) proteins, and both are associated with the Ras/PI3K/AKT pathways.

This had already been shown in the investigation of Huang et al. that Ras/PI3K/AKT pathways were turned on in the patient (50%) treated with radiotherapy in corporation with RASSF1A/RASSF2A gene silencing due to methylation. Another study from Imai T et al. investigated that RASSF2 methylated in 26% of OSCC [52, 56].

13. miRNA: Recent study has accounted that MicroRNAs have a significant role in controlling the expression of genes and are inextricably linked to initiating malignancies. Furthermore, miRNA has a variety of crucial participation in cellular processes, for instance, differentiation, proliferation, and apoptosis in both normal and abnormal cells. According to Global miR, investigators have found that overexpression of miR345, miR21, and miR181b have a vital role in malignant transformation. Several investigations have been performed to evaluate the nature of miRNA responsible for carcinoma. For example, Wong TS found that the presence of miR-133a and miR-133B was remarkably declined in OSCC. Diminishing of these levels resulted in the activation of a potential oncogene named pyruvate kinase type M2. In a separate investigation conducted by Kozaki et al., alterations in the expression levels of miR-137 and miR-193a were observed in OSCC (Oral Squamous Cell Carcinoma) cell lines. These changes in miRNA expression were associated with the epigenetic silencing of miRNAs caused by DNA hypermethylation. Such regulatory processes exert a significant influence on the development and progression of oral cancer, contributing to its pathogenesis [51, 57].

14. PTEN: PTEN has a significant role in OSCC pathogenesis, which is named downregulation, usually occurred due to hypermethylation. Although there still is a dilemma regarding the key role of PTEN in OSCC. Here, several studies have performed, such as Squarize et al. documented no aggressive nature of OSCC was found due to PTEN, whereas Shin et al. revealed that the genetic inactivation of PTEN was documented to induce OSCC carcinoma. On the other hand, various research has performed and documented that the abnormal expression of PTEN is directly related to generating and developing OSCC [37, 58].

15. E-Cadherin: Several researches have revealed that the expression of E-cadherin exists in various carcinomas. The significant role of this molecule has also been documented during both tumor generation, as well as development. However, the absence of E-cadherin has a correlation in alterations in cell functions and motility. Furthermore, the loss of its expression is also associated with tissue metastasis. Needless to say, it has found already that lower expression of E-cadherin could result in tremendous aggressive behavior during OSCC [56].

16. DAPK1: DAPK1 (death-associated protein kinase 1) gene maps on chromosome 9q34. Usually, DAPK promotes hypermethylation and causes cancer [51].

17. The impact of immune infiltration: Numerous research studies have shed light on the impact of the immune system in orchestrating adverse changes in OPMDs (Oral Potentially Malignant Disorders). These investigations reveal intriguing findings: there is a prevention of increased infiltration of T-cell CD4+ and CD8+ cells, while the expression of cell processes death protein 1 (PD-1)

and programmed cell death ligand (PDL)-1 is evident. Additionally, reduced levels of T cell CD3+ and augmented T helper 1 infiltration play significant roles in the development and progression of cancer associated with OPMDs [52].

13. Microsatellite Instability (MSI): MSI refers to the accumulation of errors in DNA microsatellites, repetitive DNA sequences found throughout the genome. Addition or deletion of microsatellite base pairs, known as microsatellite instability (MSI), is another cytogenetic feature shared by precancerous cells and OSCC [54]. In potentially malignant disorders, MSI can arise due to alterations in DNA mismatch repair genes. MSI can lead to the inactivation of tumor suppressor genes and the activation of oncogenes, contributing to cancer progression [50].

7. Diagnostic aids in detection of the potentially malignant disorders of the oral and maxillofacial region

Development of cancers and cancer-related death and negative health impacts can be reduced effectively and efficiently by early identification and treatment of premalignant disorders that have a high risk of developing into cancers [29]. This proactive strategy not only helps in detecting cancer at its earliest stages but also allows for interventions that can prevent or delay the progression of these lesions into invasive cancer. Oral lesions are early indications of oral cancer and lesions exhibiting dysplastic characteristics fall under the classification of oral potentially malignant disorders (OPMDs) [22, 59]. These OPMDs include oral leukoplakia, erythroplakia, oral submucous fibrosis (OSF), proliferative verrucous leukoplakia, and oral lichen planus [22, 25, 30], which are considered to carry a significant risk of developing into malignancy [22]. Indeed, conventional oral examination (COE) alone may have limitations when it comes to diagnosing potentially malignant disorders in the oral cavity. However, in recent years, several cutting-edge techniques and advancements in technology have been introduced to overcome these limitations (**Figure 2**).

Biopsy as a histological examination remains the gold standard for diagnosing oral potentially malignant disorders (OPMDs) [25, 30]. This traditional method involves the surgical removal of a small tissue sample from the suspicious area, which is then examined under a microscope by a pathologist to determine if any abnormal cellular changes or potentially malignant features are present. However, in recent decades, more advanced and less invasive techniques in oral cytology have been proposed as alternative methods for the diagnosis and follow-up of OPMDs. Two notable modern methods are brush biopsy and microbiopsy [30].

At present time, fluorescent technologies are broadly used in potentially malignant diagnosis. Ionic imbalance, abnormal nucleus or high ratios of nucleus/cytoplasm, unnecessary infiltrated inflammation, etc. are the indications of the abnormal mucosal lesion. During the process of examining tissues under chemiluminescent illumination, those abnormal mucosal lesions seen as aceto-white, but healthy normal mucosa display a dark blue color [59]. This difference in appearance is due to variations in how light is scattered back from the tissue between normal mucosa and abnormal lesions. Autofluorescence imaging is another fluorescent technology for potentially malignant lesion diagnosis [30]. This technology uses blue light whose wavelength is 400–460 nm [58]. When blue light is applied to tissues, normal healthy tissue displays

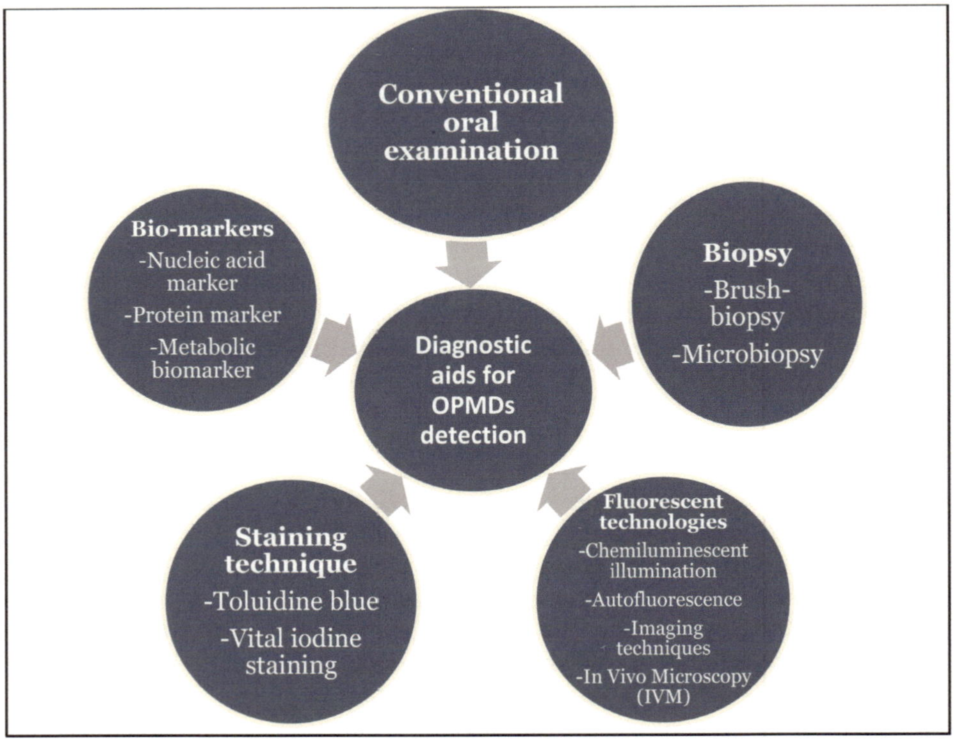

Figure 2.
Diagnostic aids for the detection of malignant transformation capacity of the potentially malignant disorders of the oral and maxillofacial region.

apple green fluorescence whose wavelength is 510 nm, while malignant lesions do not show any color (loss of autofluorescence) [59].

Imaging techniques offer additional information to physicians and can assist in planning more effective treatment strategies. In recent years, an innovative category of optical imaging technologies called *in vivo* microscopy (IVM) has emerged, offering great potential as a diagnostic tool for OPMDs. IVM works by capturing images of minute tissue characteristics through the measurement of tissue optical properties, including reflectance, scattering, absorption, and fluorescence emission. These properties often undergo changes in various disease conditions [25]. The application of IVM has shown promise in the early detection of oral potentially malignant disorders (OPMDs), although the assessment of this method is still in its preliminary stages. Multimodal imaging, optical coherence tomography, reflectance confocal microscopy, and multiphoton microscopy are some examples of IVM.

There are many clinical tissues staining technique for potentially malignant disorders diagnosis; among them toluidine blue, a cationic metachromatic dye is the most common [30]. Toluidine blue is a diagnostic adjunct that can be employed for the early detection of oral squamous cell carcinoma [60, 61]. Toluidine blue has a higher affinity for DNA and RNA, and since malignant lesions contain more DNA and RNA components, they tend to exhibit a greater uptake of the stain [60]. Toluidine blue stained areas of the dysplastic epithelium look royal blue [30]. In a study, the sensitivity of toluidine blue in detecting malignant and potentially malignant lesions was found to be 88.89% [60]. This technique not only aids in the detection of early dysplastic lesions but also assists in determining the appropriate site for biopsy [60].

Furthermore, the method is simple, noninvasive, and can be performed chair-side with a good level of diagnostic accuracy [61]. Vital iodine staining is also effective for detecting potentially malignant and malignant lesions of the oral mucosa [62].

Dysplasia is the most well-established marker but non-obligate precursors of oral squamous cell carcinoma (OSCC) [25]. Phenotypic changes in molecular markers can serve as valuable indicators for evaluating the cancer risk associated with oral potentially malignant disorders (OPMDs). These markers encompass both genetic and protein-based indicators [29]. In the context of OPMD diagnosis, stem cell self-renewal factors can be regarded as biomarkers since there are numerous shared characteristics between stem cells and cancer cells [29]. Some notable protein markers include β-catenin, cyclooxygenase 2 (COX2), c-Met, carbonic anhydrase 9 (CA9), Podoplanin, Ki-67, p16, p53, IMP3, c-Jun, SNAI1 (snail family transcriptional repressor 1), AXIN2 (axin 2), SMAD4 (SMAD family member 4), Notch1, ATM (ATM serine/threonine kinase), γH2AFX, nucleostemin (NS), SOX2 (SRY-box transcription factor 2), ALDH1, and NANOG (Nanoghomeobox), MAGE-A (MAGE family member A), cortactin, FAK (protein tyrosine kinase 2), loss of heterozygosity (LOH), Methyltransferase-Like 3 (METTL3) salivary exosomal microRNAs, etc., all of which have demonstrated potential as risk-predictive markers for OPMDs [29, 63, 64]. The abnormal alterations in the levels of these markers can be indicative of the presence of oral potentially malignant disorders (OPMDs), and the examination of those biomarkers helps to detect oral potentially malignant disorders.

Metabolic phenotypes exhibited notable distinctions between individuals with cancer and those without, indicating a promising avenue for identifying biomarkers associated with various diseases. Metabolic analysis has shown remarkable efficacy in the detection of potentially malignant disorders within the oral cavity [42]. Oral lichen planus (OLP) is classified as an oral potentially malignant disorder. A study utilizing metabolic analysis identified 21 metabolites and eight signaling pathways, which have the potential to serve as biomarkers for OLP [53]. In another study, a biomarker panel composed of Glutamic acid, LysoPE (18:0), and taurine achieved an accuracy of 87.1% in detecting OLP [42]. The researchers employed ultra-performance liquid chromatography-quadrupole/orbitrap high-resolution mass spectrometry (UHPLC-Q-Orbitrap HRMS) for this detection method.

8. Updates of therapeutic strategies in the management of the potentially malignant disorders of the oral and maxillofacial region

The management of the potentially malignant disorders of the oral and maxillofacial regions has improved dramatically in recent years, leading to new therapeutic strategies aimed at improving patient outcomes based on the size and location of it along with the overall health of the individual patient. Here are some updates in therapeutic strategies.

Surgical Excision: Surgical excision is a widely used treatment modality for the management of the potentially malignant disorders of the oropharynx and mandible. The procedure requires complete debridement of the lesion with healthy tissue margins to ensure the removal of the potentially malignant cells. Over the years, advances in surgical techniques have improved patient outcomes by reducing postoperative pain, reducing scarring, and speeding up their recovery [65]. These developments aim to further enhance the precision and outcomes of surgical excision for the potentially malignant disorders.

Minimally invasive surgical techniques have revolutionized the field by reducing the need for multiple incisions and reducing trauma to surrounding tissue. These techniques use small instruments and specialized instruments to remove the lesions accurately while preserving healthy tissue. The benefits of minimally invasive treatment include reduced postoperative pain, shorter hospital stay, faster recovery time, and improved cosmetic outcomes [66]. Such improvements have significantly improved patient experience and overall clinical outcomes.

Robotic-assisted surgery has also emerged as an alternative for the management of the potentially malignant disorders. Robotic systems provide surgeons with improved dexterity, visibility, and control during surgery. This technology allows surgeons to perform complex procedures with precision and accuracy. The use of robots in oral and maxillofacial surgery has shown promising results in terms of improved outcomes and decreased complications [67]. However, it should be noted that robot-assisted surgery may not be suitable for cases all of it, so it will require special training and knowledge.

Navigation systems have become a valuable tool in debridement surgery. These systems use advanced imaging modalities such as computed tomography (CT) scans and three-dimensional (3D) images to provide real-time guidance to surgeons during surgery through the tumor location on the precision of its relationship to the surroundings. The integration of navigation systems into surgical workflows has improved surgical outcomes and patient safety, helping surgeons to make more accurate incisions and to reduce damage to vital physiological markers [68].

Where potentially malignant disorders are extensive or there are deep lymph nodes, extensive surgery may be necessary. These procedures may require the removal of a large portion of the affected area, including the removal of adjacent structures such as bone or tissue. After excision, restorative techniques are often used to restore esthetics and function. Recent advances in reconstructive surgery, such as the use of tissue engineering and regenerative techniques, have made it easier to restore complex defects with displaced consequences effectively. These techniques use biomaterials, grafts, and tissue grafts to replace removed tissue and promote optimal healing and functional recovery [69].

While surgical excision remains an effective treatment strategy for the potentially malignant disorders, it is important to note that it is not without potential risks and complications. The extent of these risks depends on various factors, including the stage and location of the lesion, the surgical skill and experience of the surgeon, and the overall health status of the patient [70]. Potential complications can include bleeding, infection, damage to adjacent structures, and functional impairment, particularly in cases involving more extensive surgical procedures. It is crucial for surgeons to carefully assess each patient and tailor the surgical approach to minimize the risks and optimize treatment outcomes [45].

Photodynamic Therapy (PDT): To reduce the risk of recurrence and progression, adjuvant therapies are often employed in conjunction with surgical excision. These therapies aim to target any remaining abnormal cells or reduce the risk of recurrence. One such approach is the use of photodynamic therapy (PDT) following surgical excision. PDT involves the administration of a photosensitizing agent, which selectively accumulates in precancerous cells. When exposed to a specific wavelength of light, the activated agent generates reactive oxygen species, leading to the destruction of precancerous cells [71]. Combining surgical excision with PDT offers a comprehensive treatment approach that targets both the bulk of the lesion and any remaining abnormal cells, thereby improving treatment outcomes [72].

Laser Surgery: In recent years, the field of oral potentially malignant disorders management has witnessed advancements in nonsurgical treatment options as well. Laser surgery has emerged as a popular alternative to surgical excision. This technique utilizes focused beams of light to precisely target and remove potentially malignant disorders while minimizing damage to surrounding healthy tissues. Laser surgery offers advantages such as improved cosmetic outcomes, reduced postoperative discomfort, and selective destruction of precancerous cells [73]. The choice of laser type and wavelength depends on the characteristics of the lesion and the desired treatment outcome. Furthermore, laser surgery can be combined with photodynamic therapy to enhance its effectiveness through synergistic effects [74].

Topical Medications: Topical medications, such as retinoids and 5-fluorouracil (5-FU), are commonly used in the management of the potentially malignant disorders. Retinoids promote the differentiation of abnormal cells or cause cell death, helping normalize cell growth and reduce the risk of malignant transformation [75, 76]. On the other hand, 5-FU is an antimetabolite medication that interferes with DNA synthesis and inhibits the growth of abnormal cells. These topical medications have demonstrated efficacy in reducing the size and severity of lesions, preventing the development of new lesions, and improving overall outcomes. They are generally well-tolerated, with mild and temporary side effects such as skin irritation and redness [77].

Cryotherapy: Cryotherapy, utilizing extreme cold to freeze and eliminate abnormal cells, is an effective treatment for the potentially malignant disorders such as actinic cheilitis, leukoplakia, and verrucous carcinoma. It offers high cure rates and tissue healing promotion through angiogenesis and collagen formation. Successful outcomes depend on factors, such as lesion characteristics and clinician expertise. While generally safe, cryotherapy may cause side effects such as pain, blistering, and scarring [78]. Therefore, careful patient selection and postoperative care are crucial for minimizing adverse effects and optimizing treatment results. Cryotherapy serves as a valuable therapeutic option in the management of the potentially malignant disorders.

Chemoprevention: Chemoprevention aims to intervene at an early stage and halt or reverse the progression of the potentially malignant disorders, which is another area of advancement in the field. Retinoids have shown promise in preventing the progression of the potentially malignant disorders to oral cancer [79]. Nonsteroidal anti-inflammatory drugs (NSAIDs), such as celecoxib, have demonstrated anti-inflammatory and anticancer properties. Natural compounds, including green tea polyphenols such as epigallocatechin-3-gallate (EGCG), have also been investigated for their chemopreventive effects [80]. While the effectiveness of chemopreventive agents may vary depending on the specific lesion type and stage, they hold potential for reducing the burden of potentially malignant disorders and preventing their transformation into oral cancer.

In conclusion, surgical excision, along with other treatment modalities such as cryotherapy, remains a primary and effective treatment for the potentially malignant disorders of the oral and maxillofacial region. Recent advancements in surgical techniques, imaging, and diagnostics have improved precision, patient experiences, and monitoring. Adjuvant therapies such as photodynamic therapy, combined with surgical excision and cryotherapy, contribute to comprehensive lesion management. Nonsurgical options, such as laser surgery and topical medications, provide alternative approaches with cosmetic benefits. Furthermore, advancements in chemoprevention hold promise in preventing lesion progression. Interdisciplinary collaboration and ongoing research are essential for refining strategies and optimizing outcomes in the management of the potentially malignant disorders.

9. Conclusion

The genetic revelation of the potentially malignant disorders in the oral and maxillofacial region has significant implications for early detection and prevention of oral cancers. This chapter has comprehensively explored various aspects related to the potentially malignant disorders, including their clinical characteristics, epidemiology, risk factors, genetic alterations, diagnostic aids, and therapeutic strategies.

A key finding from this chapter is the critical role of genetic factors in the malignant transformation of the potentially malignant disorders. Mutations in tumor suppressor genes or oncogenes play a pivotal role in driving the progression of these lesions toward malignancy. Understanding these genetic alterations provides valuable insights into the underlying molecular pathways involved in the development of oral cancers, enabling targeted interventions and personalized treatment approaches.

Early detection of the potentially malignant disorders is of utmost importance as they serve as warning signs and precursors to oral cancers. Timely identification and monitoring of these lesions allow healthcare professionals to intervene at an early stage, significantly improving patient outcomes. Genetic markers and molecular biomarkers have emerged as promising tools for facilitating early detection and risk assessment, enabling timely interventions and preventive measures.

Managing the potentially malignant disorders of the oral and maxillofacial region requires a multidisciplinary approach involving collaboration among maxillofacial surgeons, oncologists, pathologists, and other healthcare professionals. This ensures comprehensive evaluation, accurate diagnosis, and tailored treatment plans for each patient. By implementing updated therapeutic strategies, including nonsurgical approaches such as chemoprevention and immunotherapy, and incorporating surgical interventions, when necessary, these lesions can be effectively managed, patient outcomes can be enhanced, and the burden of oral cancers on individuals and society can be reduced.

Continued research and advancements in this field are essential for further enhancing the understanding and optimizing strategies for the prevention and management of the potentially malignant disorders in the oral and maxillofacial region. By leveraging genetic insights and adopting a multidisciplinary approach, significant progress in early detection, prevention, and treatment of oral cancers can be made, ultimately improving patient prognosis and quality of life.

Author details

Nitish Krishna Das[1], A.K.M. Shafiul Kadir[2]*, Mohammad Ullah Shemanto[3], Ety Akhter[4], Ashik Sharfaraz[4], Soumik Tripura[4], Joye Kundu[5] and Ayesha Afrose Ura[6]

1 Dental Unit, Oral and Maxillofacial Surgery Department, Shaheed Suhrawardy Medical College, Dhaka, Bangladesh

2 Quest Bangladesh Biomedical Research Center, Dhaka, Bangladesh

3 Ahsania Mission Cancer and General Hospital (AMCGH), Dhaka, Bangladesh

4 Mawlana Bhashani Science and Technology University, Tangail, Bangladesh

5 Department of Biomedical Engineering, Military Institute of Science and Technology, Dhaka, Bangladesh

6 Stony Brook University, Stony Brook, NY, USA

*Address all correspondence to: shafiul_kadir@yahoo.com

References

[1] Coussens LM, Werb Z. Inflammation and cancer. Nature. 2002;**420**(6917): 860-867

[2] Ferlay J, Colombet M, Soerjomataram I, Parkin DM, Piñeros M, Znaor A, et al. Cancer statistics for the year 2020: An overview. International Journal of Cancer. 2021;**149**(4):778-789

[3] Xia C, Dong X, Li H, Cao M, Sun D, He S, et al. Cancer statistics in China and United States, 2022: Profiles, trends, and determinants. Chinese Medical Journal. 2022;**135**(05):584-590

[4] Warnakulasuriya S. Global epidemiology of oral and oropharyngeal cancer. Oral Oncology. 2009;**45**(4–5): 309-316

[5] Inchingolo F, Santacroce L, Ballini A, Topi S, Dipalma G, Haxhirexha K, et al. Oral cancer: A historical review. International Journal of Environmental Research and Public Health. 2020;**17**(9): 3168

[6] Gupta N, Gupta R, Acharya AK, Patthi B, Goud V, Reddy S, et al. Changing trends in oral cancer-a global scenario. Nepal J Epidemiol. 2016; **6**(4):613

[7] Behl I, Calado G, Stephen Flint DDS, Galvin S, Healy C, Byrne HJ, Lyng FM. Identification of precancer oral lesions using minimally invasive sampling techniques and Raman microspectroscopy. In: Proc. SPIE 12373, Optical Biopsy XXI: Toward Real-Time Spectroscopic Imaging and Diagnosis, 1237305 (6 March 2023). San Francisco, California, United States: SPIEBiOS; 2023. DOI: 10.1117/12.2654997

[8] Jr SS, Gorsky M, Ms Dds FL. Oral leukoplakia and malignant transformation. A follow-up study of 257 patients. Cancer. 1984;**53**(3):563-568

[9] Hashibe M. Risk factors for cancer of the mouth: Tobacco, betel quid, and alcohol. In: Textb Oral Cancer Prev Diagnosis Manag. Cham: Springer; 2020. pp. 23-30

[10] Mehanna H, Beech T, Nicholson T, El-Hariry I, McConkey C, Paleri V, et al. Prevalence of human papillomavirus in oropharyngeal and nonoropharyngeal head and neck cancer—Systematic review and meta-analysis of trends by time and region. Head & Neck. 2013; **35**(5):747-755

[11] Del Prete R, Ronga L, Addati G, Magrone R, Abbasciano A, Di Carlo D, et al. A retrospective study about the impact of switching from nested pcr to multiplex real-time pcr on the distribution of the human papillomavirus (HPV) genotypes. Medicina (B Aires). 2019;**55**(8):418

[12] Williams HK. Molecular pathogenesis of oral squamous carcinoma. Molecular Pathology. 2000; **53**(4):165-172

[13] Jain A. Molecular pathogenesis of Oral squamous cell carcinoma. In: Daaboul HE, editor. Squamous Cell Carcinoma. Rijeka: IntechOpen; 2019. DOI: 10.5772/intechopen.85650

[14] Todd R, Donoff RB, Wong DT. The molecular biology of oral carcinogenesis: Toward a tumor progression model. Journal of Oral and Maxillofacial Surgery. 1997;**55**(6):613-615

[15] Lorini L, Bescós Atín C, Thavaraj S, Müller-Richter U, Alberola Ferranti M, Pamias Romero J, et al. Overview of Oral

potentially malignant disorders: From risk factors to specific therapies. Cancers (Basel). 2021;**13**(15):3696

[16] Mogedas-Vegara A, Hueto-Madrid J-A, Chimenos-Küstner E, Bescós-Atín C. Oral leukoplakia treatment with the carbon dioxide laser: A systematic review of the literature. Journal of Cranio-Maxillo-Facial Surgery. 2016; **44**(4):331-336

[17] Mogedas-Vegara A, Hueto-Madrid J-A, Chimenos-Küstner E, Bescós-Atín C. The treatment of oral leukoplakia with the CO2 laser: A retrospective study of 65 patients. Journal of Cranio-Maxillo-Facial Surgery. 2015;**43**(5):677-681

[18] Lorini L, Besc C, Thavaraj S, Müller-richter U, Ferranti MA, Romero JP, et al. Overview of Oral potentially malignant disorders: From risk factors to specific therapies. Cancers. 2021;**13**:1-16

[19] Kindler S, Samietz S, Dickel S, Mksoud M, Kocher T, Lucas C, et al. Prevalence and risk factors of potentially malignant disorders of the mucosa in the general population. Ann Anat - Anat Anzeiger. 2021;**237**:151724

[20] Jiang X, Wu J, Wang J, Huang R. Tobacco and oral squamous cell carcinoma: A review of carcinogenic pathways. Tobacco Induced Diseases. 2019;**17**:29

[21] Rahman QB, Iocca O, Kufta K, Shanti RM. Global burden of head and neck cancer. Oral and Maxillofacial Surgery Clinics of North America. 2020; **32**(3):367-375

[22] Kumari P, Debta P, Dixit A. Oral potentially malignant disorders: Etiology, pathogenesis, and transformation into Oral cancer. Frontiers in Pharmacology. 2022;**13** (April):1-24

[23] Bouaoud J, Bossi P, Elkabets M, Schmitz S, Van Kempen LC, Martinez P, et al. Unmet needs and perspectives in Oral cancer prevention. Cancers (Basel). 2022;**14**(7):1815

[24] Macpherson LMD. Raising awareness of oral cancer from a public and health professional perspective. British Dental Journal. 2018;**225**(9): 809-814

[25] Yang EC, Tan MT, Schwarz RA, Richards-Kortum RR, Gillenwater AM, Vigneswaran N. Noninvasive diagnostic adjuncts for the evaluation of potentially premalignant oral epithelial lesions: Current limitations and future directions. Oral Surgery, Oral Medicine, Oral Pathology, Oral Radiology. 2018;**125**(6): 670-681

[26] Singh A, Chauhan R, Anand K, Singh M, Das S, Sinha A. Prevalence and risk factors for oral potentially malignant disorders in Indian population. Journal of Pharmacy & Bioallied Sciences. 2021; **13**(5):S398-S401

[27] Irani S. New insights into Oral cancer—Risk factors and prevention: A review of literature. International Journal of Preventive Medicine. 2020;**8**:202

[28] Sarode G, Maniyar N, Sarode SC, Jafer M, Patil S, Awan KH. Epidemiologic aspects of oral cancer. Disease-a-Month. 2020;**66**(12):100988

[29] Crawford M, Johnson EH, Liu KYP, Poh C, Tsai RYL. On the cutting edge of Oral cancer prevention: Finding risk-predictive markers in precancerous lesions by longitudinal studies. Cell. 2022;**11**(6):1033

[30] Abati S, Bramati C, Bondi S, Lissoni A, Trimarchi M. Oral cancer and precancer: A narrative review on the relevance of early diagnosis.

International Journal of Environmental Research and Public Health. 2020; **17**(24):1-14

[31] Acharya S, Singh S, Bhatia SK. Association between smokeless tobacco and risk of malignant and premalignant conditions of oral cavity: A systematic review of Indian literature. Journal of Oral and Maxillofacial Pathology. 2021; **25**(2):371

[32] Newman TM, Krishnan LP, Lee J, Adami GR. Microbiomic differences at cancer-prone oral mucosa sites with marijuana usage. Scientific Reports. 2019;**9**(1):1-8

[33] Roy D, Datta B, Pathak B. Prevalence of Oral precancerous lesions in tobacco and areca nut habituated patients in Barpeta District, Assam, India: A cross-sectional study. Journal of Clinical and Diagnostic Research. 2022;**13**:21-26

[34] Pietrobon G, Tagliabue M, Stringa LM, De Berardinis R, Chu F, Zocchi J, et al. Leukoplakia in the oral cavity and oral microbiota: A comprehensive review. Cancers (Basel). 2021;**13**(17):4439

[35] London NR, Windon MJ, Amanian A, Zamuner FT, Bishop J, Fakhry C, et al. Evaluation of the incidence of human papillomavirus-associated squamous cell carcinoma of the Sinonasal tract among US adults. JAMA Network Open. 2023;**6**(2): e2255971

[36] Katirachi SK, Grønlund MP, Jakobsen KK, Grønhøj C, von Buchwald C. The prevalence of HPV in Oral cavity squamous cell carcinoma. Viruses. 2023;**15**(2):451

[37] Huang K, Ni G, Ma B, Chen S, Luan Y, Wang T, et al. Importance of human papillomavirus infection in squamous cell carcinomas of the tongue in Guangdong Province, China. The Journal of International Medical Research. Jan 2020;**48**(1):1-13. 300060519897187. DOI 10.1177/ 0300060519897187. PMID: 31948314; PMCID: PMC7113701

[38] Dean D, Sroussi H. Oral chronic graft-versus-host disease. Frontiers in Oral Health. 2022;**3**(May):1-9

[39] de Lanna CA, da Silva BNM, de Melo AC, Bonamino MH, Alves LDB, Pinto LFR, et al. Oral lichen planus and Oral squamous cell carcinoma share key oncogenic signatures. Scientific Reports. 2022;**12**(1):1-12

[40] Rodríguez-Molinero J, Migueláñez-Medrán BDC, Puente-Gutiérrez C, Delgado-Somolinos E, Carreras-Presas CM, Fernández-Farhall J, et al. Association between oral cancer and diet: An update. Nutrients. 2021;**13**(4):1-15

[41] Bell EB, Reis IM, Cohen ER, Almuhaimid T, Smith DH, Alotaibi F, et al. Green salad intake is associated with improved oral cancer survival and lower soluble CD44 levels. Nutrients. 2021;**13**(2):1-14

[42] Wang XS, Sun Z, Liu LW, Du QZ, Liu ZS, Yang YJ, et al. Potential metabolic biomarkers for early detection of Oral lichen planus, a precancerous lesion. Frontiers in Pharmacology. 2020; **11**(November):1-10

[43] Shahi Y, Samadi FM, Mukherjee S. Plasma lipid peroxidation and antioxidant status in patients with oral precancerous lesions and oral cancer. Oral Science International. 2020;**17**(2): 86-93

[44] Wolfer S, Kunzler A, Foos T, Ernst C, Leha A, Schultze-Mosgau S. Gender and risk-taking behaviors

influence the clinical presentation of oral squamous cell carcinoma. Clinical and Experimental Dental Research. 2022; **8**(1):141-151

[45] Krishnan B, Parida S. Preoperative Evaluation and Investigations for Maxillofacial Surgery. In: Bonanthaya K, Panneerselvam E, Manuel S, Kumar VV, Rai A. editors. Oral and Maxillofacial Surgery for the Clinician. Singapore: Springer; 2021. DOI: 10.1007/978-981-15-1346-6_2

[46] George A, Sreenivasan BS, Sunil S, Varghese SS, Thomas J, Gopakumar D, Mani V. Potentially malignant disorders of oral cavity. Oral MaxillofacPathol J, 2011;**2**(1):95-100

[47] Palla S, Rangdhol V, Jeelani S, Vandana S. Determinants Of Malignant Transformation Of Oral Potentially Malignant Disorders – Covering The Gaps. Journal of Scientific Dentistry. 2015;**5**(1):36-45

[48] Canto AM, Müller H, Freitas RR, PSS S. Oral lichen planus (OLP): Clinical and complementary diagnosis. Anais Brasileiros de Dermatologia. 2010;**85**(5): 669-675

[49] Bray F, Ferlay J, Soerjomataram I, Siegel RL, Torre LA, Jemal A. Global cancer statistics 2018: GLOBOCAN estimates of incidence and mortality worldwide for 36 cancers in 185 countries. CA: a Cancer Journal for Clinicians. 2018;**68**(6):394-424

[50] Agha-Hosseini F, Sheykhbahaei N, Sadr Zadeh-Afshar M-S. Evaluation of potential risk factors that contribute to malignant transformation of oral lichen planus: A literature review. The Journal of Contemporary Dental Practice. 2016; **17**(8):692-701

[51] Mascolo M, Siano M, Ilardi G, Russo D, Merolla F, De Rosa G, et al. Epigenetic disregulation in oral cancer. International Journal of Molecular Sciences. 2012;**13**(2):2331-2353

[52] Robertson KD. DNA methylation and human disease. Nature Reviews. Genetics. 2005;**6**(8):597-610

[53] Yan YX, Zhao LX, Nan ZS. Metabolomics analysis of oral mucosa reveals profile perturbation in reticular oral lichen planus. Clinica Chimica Acta. 2018;**487**(August):28-32

[54] Mithani SK, Mydlarz WK, Grumbine FL, Smith IM, Califano JA. Molecular genetics of premalignant oral lesions. Oral Diseases. 2007;**13**(2):126-133

[55] Reibel J. Prognosis of oral pre-malignant lesions: Significance of clinical, histopathological, and molecular biological characteristics. Critical Reviews in Oral Biology and Medicine. 2003;**14**(1):47-62

[56] Huang K-H, Huang S-F, Chen I-H, Liao C-T, Wang H-M, Hsieh L-L. Methylation of RASSF1A, RASSF2A, and HIN-1 is associated with poor outcome after radiotherapy, but not surgery, in oral squamous cell carcinoma. Clinical Cancer Research. 2009;**15**(12):4174-4180

[57] Mavros A, Hahn M, Wieland I, Koy S, Koufaki ON, Strelocke K, et al. Infrequent genetic alterations of the tumor suppressor gene PTEN/MMAC1 in squamous cell carcinoma of the oral cavity. Journal of Oral Pathology & Medicine. 2002;**31**(5):270-276

[58] Lingen MW, Pinto A, Mendes RA, Franchini R, Czerninski R, Tilakaratne WM, et al. Genetics/epigenetics of oral premalignancy: Current status and future research. Oral Diseases. 2011;**17**:7-22

[59] Tatehara S, Satomura K. Non-invasive diagnostic system based on light

for detecting early-stage oral cancer and high-risk precancerous lesions—Potential for dentistry. Cancers (Basel). 2020;**12**(11):1-15

[60] Khan T, ul Alam S, Aamir M, Zeb S, Waris M, Hussain U, Ullah A. Usage of Toluidine Blue in Early Detection of Malignant and Premalignant Lesions. Journal of Khyber College of Dentistry. 2020;**10**(02):55-58. DOI: 10.33279/jkcd.v10i02.319

[61] Chishty MS, Sajid MAH, Younus S, Haq UU. Diagnostic accuracy of toluidine blue in early detection of oral squamous cell carcinoma. Journal of Fatima Jinnah Medical University. 2021; **15**(2):91-94

[62] Aftab A, Ata S. Diagnostic accuracy of Lugol's iodine staining in detection of safe margins of Oral squamous cell carcinoma. Health Science Journal. 2020; **14**:725

[63] Faur CI, Rotaru H, Osan C, Jurj A, Roman RC, Moldovan M, et al. Salivary exosomal microRNAs as biomarkers for head and neck cancer detection-a literature review. MaxillofacPlastReconstr Surg. Jun 30 2021;**43**(1):19. DOI: 10.1186/s40902-021-00303-9. PMID: 34191144; PMCID: PMC8245637

[64] Udompatanakorn C, Taebunpakul P. The expression of Methyltransferase-like 3 in Oral precancerous lesions and Oral squamous cell carcinoma. European Journal of Dental and Oral Health. 2022; **17**(2):349-356

[65] Zhou S, Zhang X, Liu W, Chen W. Evaluating surgical excision to prevent progression of oral precancerous lesions: Highlighting randomized controlled trials and cohort studies. Journal of Dental Sciences. 2023:1-7. DOI: 10.1016/j.jds.2023.05.033

[66] Birkeland AC, Kademani D, Moore MG, Blair EA. Practice patterns for initial management of oral leukoplakia amongst otolaryngologists and oral and maxillofacial surgeons. Oral Oncology. 2023;**139**:106341

[67] Liu H-H, Li L-J, Shi B, Xu C-W, Luo E. Robotic surgical systems in maxillofacial surgery: A review. International Journal of Oral Science. 2017;**9**(2):63-73

[68] Alauddin MS, Baharuddin AS, Mohd Ghazali MI. The modern and digital transformation of Oral health care: A mini review. Healthcare. 2021;**9**(2):118

[69] Borrelli MR, Hu MS, Longaker MT, Lorenz HP. Tissue engineering and regenerative medicine in craniofacial reconstruction and facial aesthetics. The Journal of Craniofacial Surgery. 2020; **31**(1):15-27

[70] Farah CS, Pollaers K, Frydrych A. Management of Premalignant Disease of the Oral mucosa. In: Brennan PA, Aldridge T, Dwivedi RC, editors. Premalignant Conditions of the Oral Cavity. Singapore: Springer Singapore; 2019. pp. 229-276

[71] Mosaddad SA, Namanloo RA, Aghili SS, Maskani P, Alam M, Abbasi K, et al. Photodynamic therapy in oral cancer: A review of clinical studies. Medical Oncology. 2023;**40**(3):91

[72] Kerr AR, Lodi G. Management of oral potentially malignant disorders. Oral Diseases. 2021;**27**(8):2008-2025

[73] Shaik K, Alanazi MN, Albilasi R, Albalawi BA, Alruwaili F. Lasers in maxillofacial surgery – Review of literature. Journal of Pharmacy & Bioallied Sciences. 2021;**13**(5):19

[74] Singh A, Zutshi A, et al. Application of Lasers in Treatment of Oral

Premalignant Lesions. Journal of Dental Health, Oral Disorders & Therapy. 2018; **6**(1):1-4. DOI: http://dx.doi.org/10.15226/jdodt.2018.00193

[75] Bachar SC, Shafiul Kadir AKM, Riajul Wahab SM, Al Hasan A. Heterocyclic Anti-cancer Compounds Derived from Natural Sources with their Mechanism of Action, Key Heterocyclic Cores for Smart Anticancer Drug-Design Part I (2022) 1: 1. Sharjah, UAE: Bentham Science Publishers; DOI: 10.2174/9789815040074122010004

[76] Pandey R, Mehrotra D. Retinoic acids in oral precancer: Utility and challenges. Journal of Oral and Maxillofacial Surgery, Medicine, and Pathology. 2020;**32**(6):549-555

[77] Serafini G, De Biase A, Lamazza L, Mazzucchi G, Lollobrigida M. Efficacy of topical treatments for the Management of Symptomatic Oral Lichen Planus: A systematic review. International Journal of Environmental Research and Public Health. 2023;**20**(2):1202

[78] Paul S. Cryotherapy: An alternative to surgical adjunct. Indian Journal of Forensic Medicine & Toxicology. 2020; **14**(4):8532-8536

[79] Sujir N, Priyanka G, Ahmed J, Saha A, Chhaparwal Y, Shenoy N. Oral cancer chemoprevention: A review. Acta Marisiensis - Ser Medica. 2023;**69**(1): 17-22

[80] Kadir AK, Murugappan S. Alkaloid Containing Natural Products for Therapeutic Use in Cancer. In: Pandurangan A, Anandasadagopan S, Alhumaydhi F, editors. Handbook of Research on Natural Products and Their Bioactive Compounds as Cancer Therapeutics. IGI Global. 2022. pp. 352-383. DOI: 10.4018/978-1-7998-9258-8.ch016

Section 6

Telomeres, Genomics and Technologies

Chapter 9

The Hidden Factor of Azoospermia: Telomere Effect

Semra Sümer Sertyel

Abstract

Nonobstructive azoospermia (NOA), which results from defective spermatogenesis is the absence of spermatozoa in the semen. NOA is a complex and multigenetic disorder that is caused by genetic and environmental factors. For the process of spermiogenesis to be fully completed, the functions of telomeres and their length in reproduction are crucial. In recent years, many studies have been published on how leukocyte telomere length might play an important role in the pathophysiology of azoospermia. They show that shorter leucocyte telomere length (LTL) is strongly associated with NOA and defective spermatogenesis. Telomeres preserve human gametogenesis and fertility while preventing chromosomal ends from eroding. The length of the telomere significantly determines how it functions. The proteins are unable to attach to telomeric regions and cannot perform capping at chromosomal ends once telomere shortening rises above a crucial threshold. It is important to include LTL evaluations as a precursor test in the treatment planning that can be created for azoospermic patients.

Keywords: leucocyte telomere length, telomere evaluation, male infertility, nonobstructive azoospermia, spermatogenetic evaluation, sperm telomere length, telomere evolution, azoospermia, semen, disorders

1. Introduction

1.1 Background of nonobstructive azoospermia

The World Health Organization (WHO) identifies infertility as a couple's failure to conceive despite having regular, unprotected sexual activity for a year. According to the most recent data, more than 50 million couples worldwide (about 15%) suffer from infertility [1, 2]. Male fertility problems affect about 50% of infertile couples [2–5], and 10–15% of this group are azoospermic patients [6]. A total of 40% of azoospermic patients have obstructive azoospermia (OA), and 60% have nonobstructive azoospermia. Normal spermatogenesis and endocrinological functions are present in OA patients; however, the male excurrent duct system is physiologically obstructed [7]. The NOA group is significantly more difficult and suffers from numerous genetically inherited issues. Chromosome abnormalities may involve autosomes or the sex chromosome. Several sex-specific chromosomal anomalies are more common in men who are azoospermic. Depending on the etiology, environmental factors might also be before or after testicular development [8, 9]. IVF laboratories have been developed to provide a variety of treatment choices, although it was previously unable to treat the azoospermic portion of the population. However, it is crucial that the causes and contributing aspects of this issue can be identified with great clarity. Varicocele, prior febrile illness, cryptorchidism, orchitis, chemotherapy medicines, radiation, and hypogonadotropic hypogonadism are a few of the reasons for azoospermia that have been identified [10–12] abnormal chromosomal numbers, autosomal mutations, abnormal karyotypes, or y chromosomal microdeletions [5, 13–17]. ICSI patients who are male are chromosomally defective in approximately four percent of cases [18]. The two most common chromosomal diseases, Robertsonian translocations and Klinefelter syndrome (KFS), impact between ten and 20% of azoospermic men. Y chromosomal microdeletions, the most typical cause of azoospermia, are seen in between 5 and 10% of infertile males. The azoospermic factor region (AZF) on the Y chromosome long arm (Yq) is crucial for germ cell growth and differentiation [19]. The AZF region includes the AZFa, AZFb, and AZFc gene locations [20]. Compared to AZFa 5%, AZFb 16%, or the combined 14% deletions, AZFc is the most frequently identified 60% deletion. This is due in part to AZFc's length being four times that of AZFa [21, 22]. NOA is a result of the most typical sex chromosomal aneuploidy cause, Klinefelter syndrome (KFS). A total of 40% of azoospermic males have KFS [19, 20]. The 47, XXY chromosomal structure, smaller testicles, high levels of follicle-stimulating hormone (FSH), low levels of testosterone (T), and high levels of luteinizing hormone (LH) are characteristics of KFS. The rare aneuploidy known as the 46, XX male disease affects 1 in 20,000 live births. A total of 90% of these males have an autosome, which is often recognizable there, or a translocation of the sex-determining region (SRY) from the Y chromosome to the X chromosome. The most common structural chromosomal defect is a Robertsonian translocation (RT). A total 1 in 1000 live newborns contains them. Infertile men experience reciprocal RTs nine times more frequently than healthy males [18]. On the other hand, there are still between 40% and 60% of people who are considered to be idiopathic [23–25]. Both the definition of this group's symptoms and the problem's origin are still unknown. The condition is characterized as idiopathic, and sperm maturation is impossible. Lifestyle and environmental variables can affect genetic makeup, but it is first essential to identify the mechanism by which the issue causes damage [8, 25–28]. These genetic developments relate to the identification of newly discovered genetic variations that cause spermatogenic failure and are used to diagnose male factor infertility [29]. The growing use of whole exome sequencing (WES) in infertility has sped studies on the telomere effect in recent years [30–34].

1.2 The role of telomeric length

Recent research findings suggest that some idiopathic reasons for male infertility and NOA may be related to telomeric structure [35–37]. The specialized DNA microsatellites known as telomeres are the linear eukaryotic chromosome ends that contain hexameric tandem repeats [38]. Terminologically, the terms "end" (telos) and "part" (meros) in Latin represent telomere. While working with Drosophila, Herman Muller discovered telomeres 80 years ago. The terminals of eukaryotic chromosomes include particular non-coding nucleoprotein structures. The functions of telomeres and their length in reproduction have been studied [39]. In humans, they develop from TTAGGG repeats and are connected to particular proteins. Telomeres serve as chromosomal stability structures and are complicated ribonucleoprotein structures with repetitive DNA sequences (5′-TTAGG-3′) [40, 41]. The functions of telomeres and their length in reproduction have previously been studied [42–45]. Females have longer telomeres than men, and telomere length varies widely between species [46]. Studies have revealed that telomere shortening through cell divisions causes early cell degeneration, which is connected to the pathogenesis of a number of illnesses, including male infertility, especially in NOA patients [47–55].

1.3 Mechanism of the effect of telomere length on male infertility

Telomeres play significant roles in meiosis by facilitating key development of gamete stages such as chromosomal pairing, synapsis, and crossing over [56]. Telomerase is a cell-based reverse transcriptase that keeps telomere length stable [57]. Telomeres in healthy somatic cells shorten with each mitotic division until they finally reach a threshold length that triggers aging, cell cycle arrest, and apoptosis [58]. Between somatic and spermatogenetic cells, there are three distinct telomere-specific changes that are known. First off, telomeres in sperm are not shortened with age, unlike those in somatic cells, ensuring that chromosomes are passed down through generations intact. Indeed, several studies have reported that increasing paternal age is actually associated with longer telomeres in spermatogenetic cells and in the leukocytes [59]. Second, numerous telomere-binding proteins have been uncovered in spermatogenetic cells [60]. Third, telomere connections are made as telomeres move toward the nuclear membrane during spermatogenesis [60].

A reverse transcriptase known as telomerase contains two structurally distinct subunits: the catalytic "telomerase reverse transcriptase" (TERT) and the human RNA matrix known as the "telomerase RNA component" (TERC). By lengthening the guanine-rich sequence and slowing the ongoing DNA loss during cell division, this ribonucleoprotein can reduce telomere shortening [61]. In contrast to stem, embryonic, and germ cells, somatic cells do not express telomerase [62, 63]. Telomere length is conserved and enhanced throughout spermatogenesis in male germ cells with active telomerase [39, 41, 45, 61–65]. Telomerase is present in testes from fetal to adulthood. The spermatogenetic stage (spermatogonia level compared to spermatocyte and spermatid stage) has increased telomerase activity. However, in spermatozoon, no telomerase activity was measured [46]. The role of sperm telomeres is not completely known, and recent many studies reported the relationship between male infertility and azoospermia [48, 55]. Some of these studies have been published on sperm telomere length in relation to spermatogenesis [55, 66, 67].

These studies show that sperm telomere length (STL) could affect male infertility and that it might be a potential new biomarker of sperm maturity. Many studies work on the relationship between telomere length (TL) in somatic cells, such as leukocyte telomere length (LTL) in azoospermia.

2. Methods for quantifying telomere length

2.1 Diagnosis of azoospermia

Azoospermia was defined as the absence of spermatozoa in the centrifuged pellet of two or three sequential specimens. These patients were further identified as OA or NOA by evaluating their medical history, physical examination, serum hormone levels (FSH, LH, total testerone, PRL, E2, and progesterone), karyotyping, Y chromosome microdeletion analysis, and imaging studies, evaluation of percutaneous epididymal sperm aspiration (PESA) or testicular sperm aspiration (TESA) results. Physical blockage of the male reproductive system was the definition of OA. These patients typically have normal hormone levels, indurated epididymides, and normal testicular volume. Spermatogenic dysfunction, which was used to define nonobstructive azoospermia, is characterized in these patients by aberrant hormone levels, tiny and soft testes [68].

For diagnosis of azoospermia, sequential semen samples (2 or 3) must be centrifugated after liquefication for 10 or 15 minutes at least 3.000 g, and after discarding of süpernatant, the remaining pellet (minimum 0,5 ml) must be evaluated under 400× magnification as wet preparation. At least two distinct day semen samples acquired more than 2 weeks apart should be analyzed, and evaluation should be carried out in accordance with 2010 World Health Organization standards [69, 70]. If even small amounts of sperm are found in the centrifuged material in cases where they have been labeled as cryptozoospermia, there is a chance of sperm cryopreservation for ICSI cycles. Numerous studies revealed that 35% of men who were assumed to have nonobstructive azoospermia actually had mature spermatozoa [71]. After several detailed evaluations of centrifugated sequential semen samples, if no mature sperm was found, this prognosis can be identified as azoospermia.

2.2 DNA isolation and methods for measurement of telomere length

Genomic DNA was extracted from peripheral blood leucocytes by conventional kits. The length of telomere length (LTL) was then determined using some different techniques.

2.2.1 Terminal restriction fragmentation (TRF)

The first method for determining telomere length was terminal restriction fragment (TRF) analysis, which is referred to be the "gold standard" approach with a combination of frequently cutting restriction enzymes that are unable to cut telomeric DNA because genomic DNA is digested using this method as the telomeric and subtelomeric regions lack recognition sites. By a probe designed specifically for telomeric DNA, the telomeric component is then identified by southern blotting or in-gel hybridization. The undamaged telomeres from each chromosome are then separated by agarose gel electrophoresis into groups based on their size. Depending on their size and intensity, telomeres will imprint at varying lengths [72, 73]. This benefit is crucial for the implementation of this technology because it means that it does not need expensive or specialist equipment. The advantages of this method include the opportunity to compare results with those from other studies and the provision of a kilobase size evaluation for telomere length. The use of restriction enzymes results in the containment of subtelomeric DNA that is close to the telomere, which causes the genuine telomere length to be incorrectly estimated as a flaw in this approach. The polymorphisms in these subtelomeric and telomeric regions may potentially result in incorrect findings interpretation. The results may also differ according to the restriction enzymes that were used. Other limitations of the TRF test include the need for significant amounts of DNA (micrograms) and the preference for telomere length analysis in blood samples rather than other tissue types. Because very short telomeres might not be able to sufficiently bind the probe, this approach cannot detect short telomeres on a small number of chromosomes [74].

2.2.2 Polymerase chain reaction-based techniques (qPCR, MMqPCR, and aTL qPCR)

This method was created in order to address the issue of requiring a significant amount of DNA in order to evaluate telomere lengths [50]. Quantitative PCR (qPCR), monochromatic multiplex quantitative PCR (MMqPCR), and absolute telomere length quantification are employed to carry out these processes. The DNA sequence is amplified by PCR using specially designed primers over the course of 20–40 cycles, with each run doubling the amount of the PCR product (the amplicon typically), a fluorophore is employed in qPCR to quantify the quantity of the target DNA sequence. After each turn, the amount of emitted fluorescence is evaluated [75]. Seven years later, the same study group improved this technique by completing the amplification of both the telomeric and single-copy DNA regions from the same tube. Monochrome multiplex quantitative PCR (MMqPCR) is the name of this novel method [76, 77]. The underlying qPCR-based method was further modified by the other research team [78]. This group referred to it as an absolute telomere length (aTL) qPCR method. Briefly, the procedure for using this approach is the same as for using the initial qPCR experiment.

Limitations of this method

- Differences between batches [79]
- Insufficient reference standards [80]
- Equipment for qPCR is required [81].

2.2.3 Single telomere length analysis (STELA)

These techniques do not offer information on individual telomere lengths; instead, they only offer evaluations of the specimen's typical telomere length. Given that it has been proposed that one or a few short telomeres act as a marker that causes cellular senescence or death [82].

Advantages

- Makes it possible to find telomeres that are critically short.
- Requires no viable cells
- Without the need for specific equipment

Limitations

- Only it provides details for a limited number of certain chromosomal ends
- Provides no mean telomere data
- Does not recognize ends lacking a telomere
- Not accurately detecting extended telomeres
- Intense in work

2.2.4 Quantitative fluorescence in situ hybridization (Q-FISH)

Quantitative fluorescence *in situ* hybridization (Q-FISH) of telomeric repeats is performed by evaluating metaphase or interphase nuclei after hybridization or labeling with a fluorescent (CCCTAA) three probe [83]. The substrate for Q-FISH, in contrast to the TRF and PCR-based tests, is cells. When used for Q-FISH, cells can be fresh, frozen, formalin-fixed, paraffin-embedded, or permeabilized. The Q-FISH approach can be used for telomeres of any size, not just average or little ones. Additionally, only by this method of evaluation will "telomere-free" endings be identified [84, 85]. For understanding how telomere length varies between chromosomes and how frequently telomere-free ends are associated with chromosomal instability, metaphase Q-FISH experiments have proved essential [86]. It is believed to be particularly helpful for figuring out the telomere length in unusual cells [87]. To correct some of its shortcomings, the use of interphase cells rather than metaphase chromosomes in these modifications is one of the Q-FISH advancements [88, 89].

Advantages

- Single telomere alterations can be found.
- Telomere lengths in particular cell types can be measured.
- When applied to metaphase chromosomes, it can assess the mean telomere length, signal signal-free ends, and identify individual telomeres.

Limitations

- Time-consuming
- Requires a great level of ability to assess chromosomes.
- Requires microscope (typically fluorescent)
- Unit of relative fluorescence for length
- When applied to metaphase chromosomes, it has the ability to recognize individual telomeres, measure end-to-end telomere length, and detect free.

Flow-FISH is another variation of the Q-FISH technique. This method combines flow cytometry with the hybridization of a pan-telomeric (binds to all telomeres) probe to cells in suspension rather than cells attached to slides, as is the case for metaphase and interphase. Cells in solution pass a laser one by one with the use of the technology known as flow cytometry. This method divides populations of cells based on their fluorescence emission or signals [90–92].

Advantages

- Using this technique, one can ascertain the mean "length" of a particular population of cells.
- When combined with antibodies, can offer details on particular cell types.
- Possibility of automation

Limitations

- Time-consuming
- Demands a high level of skill
- Requires flow sorting apparatus
- Relative fluorescence units are used to express length.

Primed in place telomeres can be labeled for the Q-FISH techniques using a primed *in situ* (PRINS) application. The telomeric sequences on metaphase chromosomes or interphase nuclei are labeled by PRINS using fluorescently-tagged nucleotides and PCR methods with telomeric primers [93, 94].

Advantages

- We can identify individual telomere modifications.
- Telomere lengths in particular cell types can be measured.
- When applied to metaphase chromosomes, it can assess the mean telomere length, signal signal-free ends, and identify individual telomeres.

Limitations

- Time-consuming
- Requires a great level of ability to assess chromosomes
- Needs a microscope, usually one that is fluorescent.
- Length as a unit of relative fluorescence
- Variability may be impacted by PCR efficiency.
- Requires only cells in metaphase to be mitotically active

2.2.5 Whole genome sequence

Next-generation sequencing has provided the opportunity to get genetic analysis for measuring telomere length measurement [95, 96].

Advantages

- More uniformity and breadth of coverage, including in the exome
- Commenting on clinically the results from genome-wide sequencing
- Data from whole genome sequencing is precise and detailed for identification.

Limitations

- The procedure produces a lot of data.

3. The role of TL and prognosis of NOA

In order to maintain genome integrity, recombination, and mitotic division, telomere length is crucial [97]. Telomere shortening has been linked to a variety of illnesses [98]. Numerous studies on male infertility with azoospermia have shown that the DNA of sperm and leucocytes from azoospermic person had shorter telomere lengths than that of males with viable sperm, indicating a potential role for telomere length in some unidentified male infertilities and azoospermia [96, 99–101]. The telomere size was evaluated in the blood leukocytes of azoospermic males [102]. In a study by Ferlin [103] on male infertility, LTL and SLT were found to be substantially associated [104–106].

4. Conclusion

The absence of mature spermatozoa in the pellet examination of concentrated seminal plasma after consecutive sampling at different times is called azoospermia. The azoospermia group without obstruction or duct problems after biochemical, genetic, and physical examinations is called nonobstructive azoospermia, and it has been determined that 1% of the population and ten percent of infertile men are affected by this problem [107]. Numerous infertile males have NOA that has been linked to a variety of genetic abnormalities, such as Y chromosomal microdeletions, karyotype disorders, and missense mutations in genes important in reproductive function. However, these alterations only explain around 25% of azoospermic cases, and around 75% of cases with severe spermatogenic impairment have an idiopathic origin [108, 109]. Recent researches strongly indicate that the pathogenesis of idiopathic NOA is complex and polygenic inheritance, which may affect spermatogenic function. Whereas the molecular mechanisms leading to NOA are still far from being understood, recent developments in genetic analysis have made it possible for our knowledge to increase significantly.

Although studies on this subject continue, telomere length and telomerase activity tests to be performed between indicated NOA patients and idiopathic patients will create an important database for azoospermia without any reason. That is important to fully clarify this possibility in the NOA patient group in order to decide the diagnosis of cryptozoospermia, which is seen in approximately 25–30% of azoospermic patient group patients. More clearly and most importantly, to inform the patient about their treatment.

5. Future trends

If it is possible to identify a target group with this type of telomere shortness for patients in the idiopathic NOA group, it may be possible to predict the rate of sperm finding in the posttreatment ejaculate or post-testicular operation procedure from these patient groups. However, there will be a need for studies to be carried out on this subject and the determination of target groups.

Author details

Semra Sümer Sertyel
Alife Hospital, Ankara, Turkey

*Address all correspondence to: semrasertyel@gmail.com

References

[1] Ray PF, Toure A, Metzler-Guillemain C, Mitchell MJ, Arnoult C, Coutton C. Genetic abnormalities leading to qualitative defects of sperm morphology or function. Clinical Genetics. 2017;**91**(2):217-232

[2] De Krertser DM. Male infertility. Lancet. 1997;**349**:787-789

[3] Ribas-Maynou J, Benet J. Single and double strand sperm DNA damage: Different reproductive effects on male fertility. Genes. 2019;**10**(2):105

[4] Cho CL, Agarwal A. Role of sperm DNA fragmentation in male factor infertility: A systematic review. Arab Journal of Urology. 2017;**16**(1):21-34

[5] Lee JY, Dada R, Sabanegh E, Carpi A, Agarwal A. Role of genetics in azoospermia. Urology. 2011;**77**: 598-601

[6] Practice Committee of American Society for Reproductive Medicine in collaboration with Society for Male Reproduction and Urology. The management of infertility due to obstructive azoospermia. Fertility and Sterility. 2008;**90**:S121-S124

[7] Hamada AJ, Esteves SC, Agarwal A. A comprehensive review of genetics and genetic testing in azoospermia. Clinics. 2013;**68**(Suppl 1):39-60

[8] Ferlin A, Arredi B, Foresta C. Genetic causes of male infertility. Reproductive Toxicology. 2006;**22**:133-141

[9] Jarow JP, Espeland MA, Lipshultz LI. Evaluation of the azoospermic patient. The Journal of Urology. 1989;**142**:62-65

[10] Wosnitzer M, Goldstein M, Hardy MP. Review of azoospermia. Spermatogenesis. 2014;**31**:4

[11] Berookhim BM, Schlegel PN. Azoospermia due to spermatogenic failure. Urologic Clinics of North America. 2014;**41**(1):97-113

[12] O'Flynn O'Brien KL, Varghese AC, Agarwal A. The genetic causes of male factor infertility: A review. Fertility and Sterility. 2010;**93**:1-12

[13] Bashamboo A, Ferraz-de-Souza B, Lourenco D, Lin L, Sebire NJ, Montjean D, et al. Human male infertility associated with mutations in NR5A1 encoding steroidogenic factor 1. American Journal of Human Genetics. 2010;**87**:505-512

[14] Wang XN, Li ZS, Ren Y, Jiang T, Wang YQ, Chen M, et al. The Wilms tumor gene, Wt1, is critical for mouse spermatogenesis via regulation of sertoli cell polarity and is associated with NOA in humans. PLoS Genetics. 2013;**9**:8

[15] Yang F, Silber S, Leu NA, Oates RD, Marszalek JD, Skaletsky H. TEX11 is mutated in infertile men with azoospermia and regulates genome-wide recombination rates in mouse. EMBO Molecular Medicine. 2015;**7**:1198-1210

[16] Wu H, Sun L, Wen Y, Xia L. Major spliceosome defects cause male infertility and are associated with nonobstructive azoospermia in humans. PNAS. 2016;**113**:4134-4139

[17] Bracke A, Peeters K, Punjabi U, Hoogewijs D, Dewilde S. A search for molecular mechanisms underlying male idiopathic infertility. Reproductive BioMedicine Online. 2018;**36**(3):327-339

[18] Chandley AC. The chromosomal basis of human infertility. British Medical Bulletin. 1979;**35**:181-186

[19] Ferlin A, Raicu F, Gatta V, Zuccarello D, Palka G, Foresta C. Male infertility: Role of genetic background. Reproductive Biomedicine Online. 2007;**14**:734-745

[20] Walsh TJ, Pera RR, Turek PJ. The genetics of male infertility. Seminars in Reproductive Medicine. 2009;**27**: 124-136

[21] Hopps CV, Mielnik A, Goldstein M, Palermo GD, Rosenwaks Z, Schlegel PN. Detection of sperm in men with Y chromosome microdeletions of the AZFa, AZFb and AZFc regions. Human Reproduction. 2003;**18**:1660-1665

[22] Ferlin A, Arredi B, Speltra E, Cazzadore C, Selice R, Garolla A. Molecular and clinical characterization of Y chromosome microdeletions in infertile men: A 10-year experience in Italy. The Journal of Clinical Endocrinology and Metabolism. 2007;**92**:762-770

[23] Salonia ACB, Carvalho J, Corona G. EAU Guidelines on Sexual and Reproductive Health. EAU Annual Congress Milan. Arnhem: The Netherlands: EAU Guidelines Office; 2021. pp. 106-116

[24] Lopes AC, Oliveira PF, Sousa M. Shedding light into the relevance of telomeres in human reproduction and male factor infertility. Biology of Reproduction. 2019;**100**:318-330

[25] Oates R. Evaluation of the azoospermic male. Asian Journal of Andrology. 2012;**14**(1):82-87

[26] Matzuk MM, Lamb DJ. Genetic dissection of mammalian fertility pathways. Nature Cell Biology. 2002;**4**(Suppl):s41-s49

[27] Nuti F, Krausz C. Gene polymorphisms/mutations relevant to abnormal spermatogenesis. Reproductive Biomedicine Online. 2008;**16**:504-513

[28] Poongothai J, Gopenath TS, Manonayaki S. Genetics of human male infertility. Singapore Medical Journal. 2009;**50**:336-347

[29] Krausz C, Riera-Escamilla A. Genetics of male infertility. Nature Reviews. Urology. 2018;**15**:369-384

[30] Mbango JN, Coutton C, Arnoult C. Genetic causes of male infertility: Snapshot on morphological abnormalities of the sperm flagellum. Basic and Clinical Andrology. 2019;**9**:2

[31] Oud MS, Volozonoka L, Smits RM. A systematic review and standardized clinical validity assessment of male infertility genes. Human Reproduction. 2019;**34**:932-941

[32] Kasak L, Laan M. Monogenic causes of non-obstructive azoospermia: Challenges, established knowledge, limitations and perspectives. Human Genetics. 2020;**140**(1):135-154

[33] Xavier M, Salas-Huetos A, Oud M. Disease gene discovery in male infertility: Past, present and future. Human Genetics. 2020;**140**(1):7-19

[34] O'sullivan R, Karlseder J. Telomeres: Protecting chromosome against genome instability. Molecular and Cellular Biology. 2010;**11**:171-181

[35] Thilagavathi J, Kumar M, Mishra SS, Venkatesh S, Kumar R, Dada R. Analysis of sperm telomere length in men with idiopathic infertility. Archives

of Gynecology and Obstetrics. 2013;**287**(4):803-807

[36] Yuan Y, Tan Y, Qui X, Luo H, Li Y, Li R, et al. Sperm telomere length as a novel biomarker of male infertility and embriyonic development; a systematic review and meta analysis. Frontiers in Endocrinology. 2022;**13**:1079966

[37] Lin KW, Yan J. The telomere length dynamic and methods of its assessment. Journal of Cellular and Molecular Medicine. 2005;**9**(4):977-989

[38] Thilagavathi J, Venkatesh S, Dada R. Telomere length in reproduction. Andrologia. 2013;**45**:289-304

[39] Montpetit AJ, Alhareli AA, Montpetit M, Starkweather AR, Elmore LW, Filler K, et al. Telomere length: A review of methods for measurement. Nursing Research. 2014;**63**(4):289-299

[40] Shay JW, Wright WE. Telomeres and telomerase: Implications for cancer and aging. Radiation Research. 2001;**155**:188-193

[41] Liu L, Blasco M, Trimarchi J, Keefe D. An essential role for functional telomeres in mouse germ cells during fertilization and early development. Developmental Biology. 2002;**249**:74-84

[42] Liu L, Franco S, Spyropoulos B, Moens PB, Blasco MA, Keefe DL. Irregular telomeres impair meiotic synapsis and recombination in mice. Proceedings of the National Academy of Sciences of the United States of America. 2004;**101**:6496-6501

[43] Factor-Litvak P, Susser E, Kezios K, McKeague I, Kark JD, Hoffman M, et al. Leukocyte telomere length in newborns: Implications for the role of telomeres in human disease. Pediatrics. 2016;**137**:e20153927

[44] Samassekou O, Gadji M, Drouin R, Yan J. Sizing the ends: Normal length of human telomeres. Annals of Anatomy. 2010;**192**:284-291

[45] Achi MV, Ravindranath N, Dym M. Telomere length in male germ cells is inversely correlated with telomerase activity. Biology of Reproduction. 2000;**63**:591-598

[46] Kohlrausch FB, Wang F, Chamani I, Keefe DL. Telomere shortening and fusions: A link to aneuploidy in early human embryo development. Obstetrical & Gynecological Survey. 2021;**76**:429-436

[47] Rocca MS, Foresta C, Ferlin A. Telomere length: Lights and shadows on their role in human reproduction. Biology of Reproduction. 2019;**100**:15-17

[48] Herrmann M, Pusceddu I, März W, Herrmann W. Telomere biology and age-related diseases. Clinical Chemistry and Laboratory Medicine. 2018;**56**:1210-1222

[49] Youngren K, Jeanclos E, Aviv H, Kimura M, Stock J, Hanna M, et al. Synchrony in telomere length of the human fetus. Human Genetics. 1998;**102**:640-643

[50] Ferlin A, Rampazzo E, Rocca MS, Keppel S, Frigo AC, De Rossi A, et al. In young men sperm telomere length is related to sperm number and parental age. Human Reproduction. 2013;**28**:3370-3376

[51] Vasilopoulos E, Fragkiadaki P, Kalliora C, Fragov D, Vakonaki E, Tsoukalas D, et al. The association of female and male infertility with telomere length. International Journal of Molecular Medicine. 2019;**44**(2):375-389

[52] Turner S, Hartshorne GM. Telomere lengths in human pronuclei, oocytes

and spermatozoa. Molecular Human Reproduction. 2013;**19**:510-518

[53] Antunes DM, Kalmbach KH, Wang F, Dracxler RC, Seth-Smith ML, Kramer Y, et al. A single-cell assay for telomere DNA content shows increasing telomere length heterogeneity, as well as increasing mean telomere length in human spermatozoa with advancing age. Journal of Assisted Reproduction and Genetics. 2015;**32**:1685-1690

[54] Yang Q, Zhao F, Dai S, Zhang N, Zhao W, Bai R, et al. Sperm telomere length is positively associated with the quality of early embryonic development. Human Reproduction. 2015;**30**:1876-1881

[55] Tiepolo L, Zuffardi O. Localization of factors controlling spermatogenesis in the nonfluorescent portion oi the human Y chromosome long arm. Human Genetics. 1976;**124**:119-124

[56] Blacburn EH. Structure and function of telomeres. Nature. 1991;**3501**:569-573

[57] Harley CB, Furtcher AB, Greider CW. Telomeres aging of human fibroblast. Nature. 1990;**345**:458-460

[58] Aston KI, Hunt SC, Susser E, Kimura M, Factor-Litvak D, Carrel D, et al. Divergence of sperm and leucocyte age-dependent telomere dynamics: Implication for male driven evaluation of telomere lenghts in human. Molecular Human Reproduction. 2012;**181**:517-522

[59] Gineitis AD, Zalenskaya IA, Yau PM, Bradbury EM, Zalensky AO. Human sperm telomere -binding complex involves histone H2B and secures telomere membrane attachment. The Journal of Cell Biology. 2000;**1511**:1591-1598

[60] Lindsey J, McGill NI, Lindsey LA, Green DK, Cooke HJ. In vivo loss of telomeric repeats with age in humans. Mutation Research. 1991;**256**:45-48

[61] De Lange T. Shelterin: The protein complex that shapes and safeguards human telomeres. Genes. 2005;**19**:2100

[62] Fattet AJ, Chaillot M, Koscinki I. Telomere length, a new biomarker of male (in) fertility? A systematic review of the literature. Genes. 2023;**14**(2):425

[63] Blackburn EH. Telomeres and telomerase: Their mechanisms of action and the effects of altering their functions. FEBS Letters. 2005;**579**:859-862

[64] Shay JW, Wright WE, Werbin H. Defining the molecular mechanisms of human cell immortalization. Biochimica et Biophysica Acta. 1991;**1072**:1-7

[65] Torra-Massana M, Barragán M, Bellu E, Oliva R, Rodríguez A, Vassena R. Sperm telomere length in donor samples is not related to ICSI outcome. Journal of Assisted Reproduction and Genetics. 2018;**35**:649-657

[66] Zhao F, Yang Q, Shi S, Luo X, Sun Y. Semen preparation methods and sperm telomere length: Density gradient centrifugation versus the swim up procedure. Scientific Reports. 2016;**6**:39051

[67] Eisenberg DTA, Kuzawa CW. The paternal age at conception effect on ofspering telomere length; mechanistic, comperative and adaptive perspectives. Philosophical Transactions of the Royal Society B: Biological Sciences. 2018;**5**(373):1741

[68] Fujisawa M, Tanaka H, Tatsumi N, Okada H, Arakawa S, Kamidono S. Telomere activity in the testis of infertile patients with selected causes. Human Reproduction. 1998;**13**(6):1476-1479

[69] WHO. World Health Organization: WHO Laboratory Manual for the Examination and Processing of Human Semen. 5th ed. Geneva: WHO Press; 2010

[70] Schlegel PN. Causes of azoospermia and their management. Reproduction, Fertility, and Development. 2004;**16**(5):561-572

[71] Ron-El R, Strassburger D, Friedler S, Komarovski D, Bern O, Soffer Y. Extended sperm preparation: An alternative to testicular sperm extraction in non-obstructive azoospermia. Human Reproduction. 1997;**12**(6):1222-1226

[72] Allshire RC, Dempster M, Hastie ND. Human telomeres contain at least three types of G-rich repeat distributed non-randomly. Nucleic Acids Research. 1989;**17**:4611-4627

[73] Harley CB, Futcher AB, Greider CW. Telomeres shorten during ageing of human fibroblasts. Nature. 1990;**345**:458-460

[74] Aubert G, Hills M, Lansdorp PM. Telomere length measurement—Caveats and a critical assessment of the available technologies and tools. Mutation Research/Fundamental and Molecular Mechanisms of Mutagenesis. 2012;**730**:59-67

[75] Ding C, Cantor CR. Quantitative analysis of nucleic acids: The last few years of progress. Journal of Biochemistry and Molecular Biology. 2004;**37**:1-10

[76] Cawthon RM. Telomere measurement by quantitative PCR. Nucleic Acids Research. 2002;**30**(10):e47

[77] Cawthon RM. Telomere length measurement by a novel monochrome multiplex quantitative PCR method. Nucleic Acids Research. 2009;**37**:3

[78] O'Callaghan NJ, Fenech M. A quantitative PCR method for measuring absolute telomere length. Biological Procedures Online. 2011;**13**:3

[79] Cunningham JM, Johnson RA, Litzelman K, Skinner HG, Seo S, Engelman CD, et al. Telomere length varies by DNA extraction method: Implications for epidemiologic research. Cancer Epidemiology, Biomarkers and Prevention. 2013;**22**:2047-2054

[80] Koppelstaetter C, Jennings P, Hochegger K, Perco P, Ischia R, Karkoszka H, et al. Effect of tissue fixatives on telomere length determination by quantitative PCR. Mechanisms of Ageing and Development. 2005;**126**:1331-1333

[81] Aviv A, Hunt SC, Lin J, Cao X, Kimura M, Blackburn E. Impartial comparative analysis of measurement of leukocyte telomere length/DNA content by southern blots and qPCR. Nucleic Acids Research. 2011;**39**:20

[82] Abdallah P, Luciano P, Runge KW, Lisby M, Géli V, Gilson E, et al. A two-step model for senescence triggered by a single critically short telomere. Nature Cell Biology. 2009;**11**:988-993

[83] Kimura M, Stone RC, Hunt SC, Skurnick J, Lu X, Cao X, et al. Measurement of telomere length by the southern blot analysis of terminal restriction fragment lengths. Nature Protocols. 2010;**5**:1596-1607

[84] Lansdorp PM, Verwoerd NP, van de Rijke FM, Dragowska V, Little M-T, Dirks RW, et al. Heterogeneity in telomere length of human chromosomes. Human Molecular Genetics. 1996;**5**:685-691

[85] Krejci K, Koch J. Improved detection and comparative sizing of human chromosomal telomeres *in situ*. Chromosoma. 1998;**107**:198-203

[86] Artandi SE, Chang S, Lee S-L, Alson S, Gottlieb GJ, Chin L, et al. Telomere dysfunction promotes non-reciprocal translocations and epithelial cancers in mice. Nature. 2000;**406**:641-645

[87] Goldman FD, Aubert G, Klingelhutz AJ, Hills M, Cooper SR, Hamilton WS, et al. Characterization of primitive hematopoietic cells from patients with dyskeratosis congenita. Blood. 2008;**111**:4523-4531

[88] Nettle D, Gadalla SM, Lai TP, Susser E, Bateson M, Aviv A. Measurement of telomere length for longitudinal analysis; implications of assay precision epidemiology. American Journal of Epidemiology. 2021;**190**(7):1406-1413

[89] Vera E, Blasco MA. Beyond average: Potential for measurement of short telomeres. Aging. 2012;**4**:379-392

[90] Hultdin M, Grōnlund E, Norrack K-F, Eriksson-Lindstrōm E, Roos G, Just T. Telomere analysis by fluorescence in situ hybridization and flow cytometry. Nucleic Acids Research. 1998;**26**:3651-3656

[91] Alter BP, Baerlocher GM, Savage SA, Chanock SJ, Weksler BB, Willner JP, et al. Very short telomere length by flow fluorescence in situ hybridization identifies patients with dyskeratosis congenita. Blood. 2007;**110**:1439-1447

[92] Darmishonnejad Z, Kheirabadi F, Tavalaee M, Kheirabadi M, Zohrabi D, Nasr-Efehani M. Releationship between sperm telomere length and sperm quality in infertile man. Andrologia. 2020;**52**(5):e13546

[93] Therkelsen AJ, Nielsen A, Koch J, Hindkjaer J, Kølvraa S. Staining of human telomeres with primed in situ labeling (PRINS). Cytogenetic and Genome Research. 1995;**68**:115-118

[94] Hjelmborg JB, Dalgerd C, Moller S, Steenstrup T, Kimura M, Christengen K, et al. The heritability of leucocyte telomere length daynamics. Journal of Medical Genetics. 2014;**52**(5):297-302

[95] Ding Z, Mangino M, Aviv A, Spector T, Durbin R. Estimating telomere length from whole genome sequence data. Nucleic Acids Research. 2014;**42**(9):e75

[96] Salas-Huetos A, Tuttelmann FJ, Wyrwoll M, Kliesch SM, Lopez A, Concalves J, et al. Disruption of human meiotic telomere complex genes TERB1, TERB2 and MAJIN in men with non-obstructive azoospermia. Human Genetics. 2021;**140**(1):217-227

[97] Cheng EU, Chen SU, Lee TH, Pai YP, Huang LS, Huang CC, et al. Evaluation of telomere length in cumuluscells as a potential biomarker of oocyte and embryo quality. Human Reproduction. 2013;**28**(4):929-936

[98] Calado R, Young N. Telomeres in disease. F1000 Medicine Reports. 2012;**4**:8

[99] Fice HEF, Robeire B. Telomere dynamics throughout spermatogenesis. Genes. 2019;**10**(7):525

[100] Heidery H, Pouresmaeili F, Mirfakhraie R, Omrani M, Ghaedi H, Fazeli Z, et al. An association study between longitudinal changes of leukocyte telomere and the risk of azoospermia in a population of Iranian

infertile men. Iranian Biomedical Journal. 2018;**22**(4):231-236

[101] Yang Q, Luo X, Bai R, Zhao F, Dai S, Li F, et al. Shorter leukocyte telomere length is associated with risk of nonobstructive azoospermia. Fertility and Sterility. 2018;**110**(4):648-654.e1

[102] Cioppi F, Rosta V, Krausz C. Genetics of Azoospeermia. International Journal of Molecular Sciences. 2021;**22**(6):3264

[103] Ferlin A, Rampazzo E, Rocca MS, Keppel S, Frigo AC, De Rossi A, et al. In young men sperm telomere lengh is releated to sperm number and parental age. Human Reproduction. 2013;**28**(12):3370-3376

[104] Boniewska-Bernacka E, Pańczyszyn A, Cybulska N. Telomeres as a molecular marker of male infertility. Human Fertility. 2019;**22**:78-87

[105] Balmori C, Varela E. Should we consider telomere length and telomerase activity in male Factor infertility? Current Opinion in Obstetrics & Gynecology. 2018;**30**:197-202

[106] Arsenis NC, You T, Ogawa EF, Tinsley GM, Ziu L. Physical activity and telomere length: Impact of aging and potential mechanisms of action. Oncotarget. 2017;**8**(27):45008-45019

[107] Maduro MR, Lamb DJ. Understanding new genetics of male infertility. The Journal of Urology. 2002;**168**(21):2197-2203

[108] Punab M, Poolamets O, Paju P, Vihljajev V, Pomm K, Ladva R. Causes of male infertility: A 9- year prospective monocentre study on 1737 patients with reduced total sperm counts. Human Reproduction. 2017;**32**:18-31

[109] Cerván-Martín M, Castilla JA, Palomino-Morales RJ, Carmona FD. Genetic landscape of nonobstructive azoospermia and new perspectives for the clinic. Journal of Clinical Medicine. 2020;**9**:300

Chapter 10

Current Technologies for Measuring or Predicting Telomere Length from Genomic Datasets

Ting Zhai and Zachary D. Nagel

Abstract

The gold standard for measuring telomere length is technically challenging, which limits its use in large population studies. Numerous bioinformatics tools have recently been developed to estimate telomere length using high-throughput sequencing data. This allows for scaling up telomere length estimates in large datasets. Telomere length depends substantially on genetics, and many genetic studies have looked at this relationship, which provides an opportunity to predict telomere length from genotyping data. However, in part because environment also significantly affects telomere length, the accuracy of telomere length predictions and estimates made from genomic data remains uncertain. In this chapter, we will summarize currently available bioinformatics tools for predicting or measuring telomere length from genomics datasets, and we will discuss each method's limitations and advantages.

Keywords: DNA sequencing, genomics, genetics, population sciences, bioinformatics

1. Introduction

Telomere length, as a biomarker of aging and genome integrity, has frequently been used in aging and health research. Telomere length in blood leukocytes or peripheral blood mononuclear cells (PBMCs) reflects the progressive shortening of telomeres in hematopoietic stem and progenitor cells (HSPCs) and correlates with telomere length in other tissues [1]. Among leukocyte types, naïve T-cells and B-cells have the longest telomeres, whereas NK- and memory T-cells have shorter telomeres within the same individual [2]. However, consistent with blood telomere length as a representative marker for telomere attrition in other tissues, telomere length in all cell types shows an inverse correlation with participant age. Many age related diseases are associated with diminished telomere length in leukocytes or PBMCs, including cardiovascular disease [3] and cancer [4]. Accurate and accessible telomere length techniques are increasingly needed in population studies to evaluate the biological aging process and genome maintenance, and may represent a target for the development of strategies for early detection and prevention of disease.

The gold standard measurement of telomere length is by Terminal Restriction Fragment (TRF) analysis. This approach involves frequent-cutting restriction enzymes that do not recognize the telomeric repeats, followed by Southern hybridization [5]. The entire process of TRF can take up to one week, making it infeasible to apply on a large population scale. Flow cytometry-based fluorescent in situ hybridization (flow-FISH) is frequently used in clinical settings for patient diagnoses of telomere related disease [6]. An advantage of this approach is that it allows for determination of telomere length for specific cell population, but it requires expertise in flow cytometry and still labor intensive. Currently, the monochrome multiplex–quantitative polymerase chain reaction (MM–qPCR) is widely used in population-based studies because it is less labor intensive and relevantly fast in the turnaround time [7]. However, even this approach requires careful attention to details that may affect the results of the analysis and places demands on sample processing that may not be compatible with population studies [8].

Recently, genomics datasets have been produced from many large population cohorts, enabling new approaches to estimating telomere length through bioinformatics tools. These approaches afford an opportunity to estimate telomere length in large-scale population datasets. In this chapter, we summarize currently available tools for predicting or measuring telomere length from genomics datasets, and we will discuss each method's limitations and advantages.

2. Predicting telomere length from genotyping datasets

2.1 Genome-wide association studies of telomere length

Telomere length variation between individuals can be explained at least in part by genetics. Telomeropathies are Mendelian diseases characterized by impaired telomere maintenance and caused by defects in genes involved in telomere maintenance [9]. Genome-Wide Association Studies (GWAS) of leukocyte telomere length have consistently identified genetic loci within genes with key roles in telomere length regulation. These studies have shed light on the connection between genetic variants and telomere length, allowing for the prediction of telomere length using an individual's genetic information.

In 2021, Codd et al. reported the largest GWAS on telomere length to date, using genetic data from over 472,000 participants in the UK Biobank [10]. The study identified 197 independent genetic variants associated with leukocyte telomere length at 138 genomic loci, with 108 being newly discovered. Genes involved in regulating telomeres were found in 44 loci, including those that encode components of the Shelterin and CTC1-STN1-TEN1 (CST) complexes. The newly discovered loci also included genes involved in the alternative lengthening of telomeres (ALT) pathway and factors that modify key telomere proteins post-translationally. Additionally, genes that regulate telomerase such as TERC and TERT were reported. This study further established a method for predicting telomere length from the identified loci and revealed its relationship to age-related disease outcomes involving multiple biological traits and chronic pathologies.

Also in 2021, Chang et al. reported the largest GWAS of leukocyte telomere length in an Asian population, among 25,533 Chinese Singaporean individuals [11]. The study identified three variants in or near the POT1, TERF1 and STN1 genes that were associated with telomere length and specific to East Asians. Additionally, the authors reported a significantly increased risk of incident lung cancer with increased genetic telomere length. Upon further analysis stratifying on subtypes, this association was found only in lung adenocarcinoma.

These studies provide valuable insights into the genetic determinants of telomere length and its association with various health outcomes. Further research is needed to fully understand the mechanisms underlying these associations and to develop potential interventions to improve health outcomes.

2.2 Predicting telomere length using Mendelian randomization

In this section, we discuss the emerging use of Mendelian randomization (MR) to predict telomere length from germline genotyping data. MR is an approach that uses genetic variants to infer causal relationships between genotype and phenotype. It is based on Mendel's laws of inheritance and causal inference theory, using instrumental variables to account for unmeasured confounding [12]. MR has recently gained popularity, with a growing number of methodologies and applied studies being published, enabled by an increasing availability of genetic data [13].

There are three basic assumptions of MR in telomere length studies: (1) the relevance assumption - the genetic variant(s) used as instruments must be strongly associated with telomere length; (2) the independence assumption - the genetic variant(s) should be independent of the outcome, given telomere length and all confounders; and (3) the exclusion assumption - the genetic variant(s) should only influence the outcome through their effect on telomere length. The first assumption is directly related to our topic of predicting telomere length from germline variants.

To identify germline variants that can be used as instrumental variables for telomere length, a set of screening and filtering criteria can be applied to available genetic datasets. In this section, we outline our approach for selecting genetic instruments for telomere length in European populations. We have chosen to focus on European populations because they currently have the largest available databases among all populations in published genetic datasets.

In January 2023, we conducted a search of telomere length-related studies on the GWAS Catalog (https://www.ebi.ac.uk/gwas/efotraits/EFO_0004505), and identified 303 SNPs from 20 published GWAS studies. A detailed flowchart has been provided to illustrate each step of our screening and filtering process (**Figure 1**). We then systematically screened these studies using the following exclusion criteria: (1) non-European ethnicity; (2) telomere length measured in patient populations; (3) telomere length measured in cell types other than leukocytes or PBMCs; and (4) inconsistent units of telomere length measurement. After applying these criteria, 9 out of the 20 studies were selected for full-text screening.

During the full-text screening process, we eliminated 4 out of the 9 studies because they were conducted earlier with smaller sample sizes and shared the same populations as the included studies. We also expanded our list of potential genetic instrument candidates by adding more SNPs from the study results and supplementary lists along with the published manuscript. This resulted in a candidate list of 143 SNPs from 5 studies [14–18], which was reduced to a final list of 138 SNPs after meta-analyzing for duplicated SNPs.

From this final candidate list, we applied SNP-specific filtering criteria to identify strong SNPs for telomere length: (1) genome-wide significance with p-value $<5\times10^{-8}$, as defined by the relevance assumption; (2) minor allele frequency (MAF) > 0.01, to avoid potential statistical bias from SNPs with low MAF; and (3) pruning for linkage disequilibrium (LD) at an R^2 coefficient of correlation <0.01, to ensure that the included genetic instruments are independent of each other, and that we have only one representative SNP per region of LD.

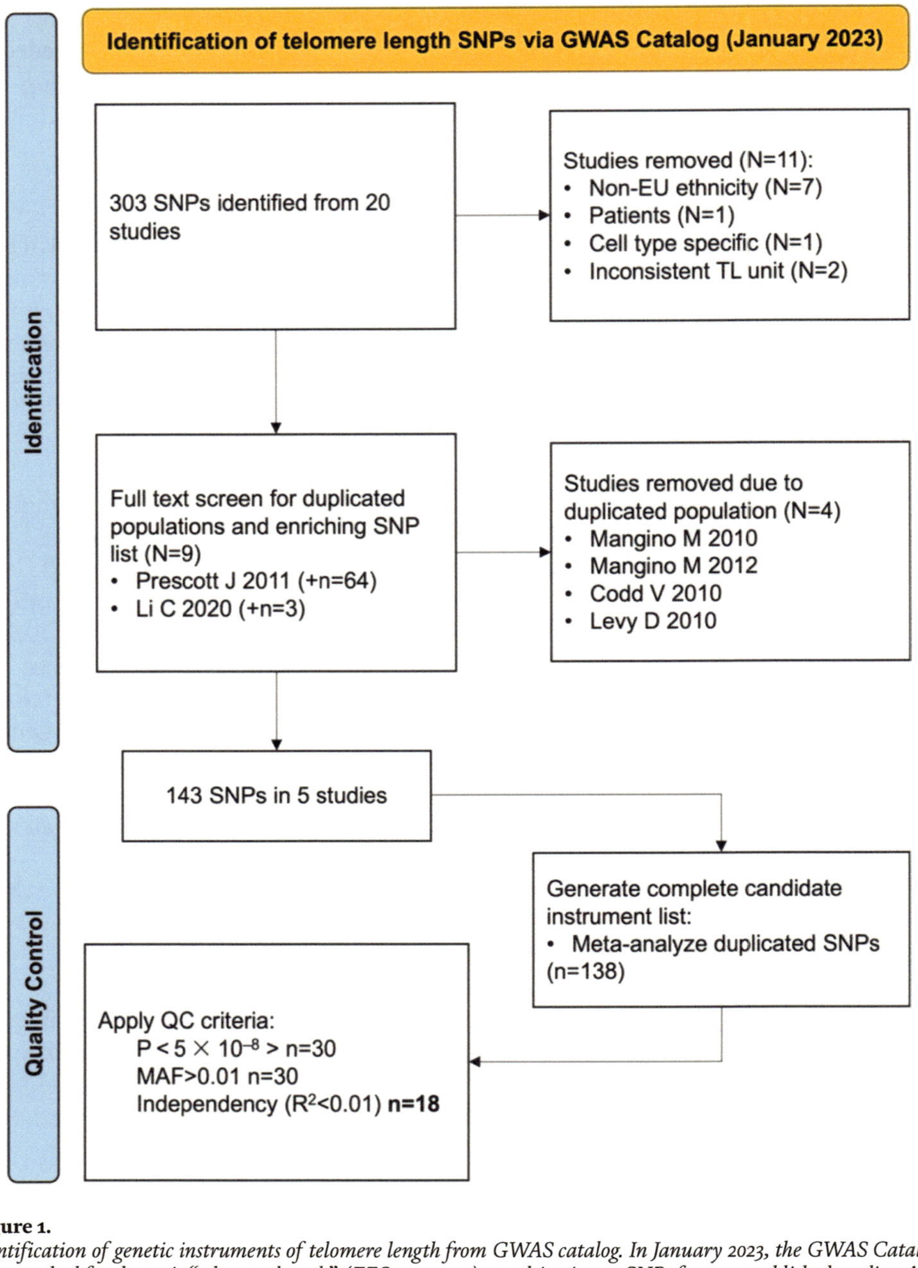

Figure 1.
Identification of genetic instruments of telomere length from GWAS catalog. In January 2023, the GWAS Catalog was searched for the trait "telomere length" (EFO_0004505), resulting in 303 SNPs from 20 published studies. After screening the abstracts, seven studies were removed due to non-European populations, one due to being conducted on patients, one due to reporting cell-type specific telomere length, and two due to reporting associations in different units. Of the remaining nine studies, four were removed after full-text screening because they used the same study populations as other studies but had smaller sample sizes. This left five studies with a total of 143 SNPs. After meta-analysis to remove duplicate SNPs and SNP-specific filtering, 30 SNPs had genome-wide significance ($P < 5 \times 10^{-8}$) and minor allele frequency > 0.01. Of these, 18 were independent ($R^2 < 0.01$) within a 10 Mb region. The final list consisted of 18 independent SNPs as genetic instruments for telomere length, eligible for Mendelian randomization analysis. Abbreviation: SNP, single-nucleotide polymorphism; GWAS, genome-wide association study; EU, European; TL, telomere length; QC, quality control; MAF, minor allele frequency. N denotes the sample size for studies, and n denotes the sample size for SNPs.

After applying these criteria, we identified 30 SNPs with genome-wide significant p value $<5\times10^{-8}$ and MAF > 0.01. We then performed LD clumping with $R^2 < 0.01$ to identify the top SNP per 10 Mb region, resulting in a total of 18 SNPs as our final genetic instruments for telomere length (**Table 1**).

With our final genetic instruments for predicting telomere length through genetic variants in hand, we can use these SNPs to conduct Mendelian randomization analyses. This approach allows us to infer causal relationships between telomere length and other phenotypes by using the genetic instruments as proxies for telomere length.

CHR	POS	GENE	SNPS	PVAL	BETA	SE	REF	EFT
1	226,562,621	*PARP1*	rs3219104	9.31E-11	0.0417	0.0064	A	C
2	54,248,729	ACYP2	rs11125529	8.00E-10	−0.0560	0.0100	A	C
3	169,514,585	*LRRC34 (TERC)*	rs10936600	6.42E-51	−0.0858	0.0057	A	T
3	101,232,093	*SENP7*	rs55749605	2.38E-08	−0.0373	0.0067	C	A
4	164,048,199	*NAF1*	rs4691895	1.47E-21	0.0577	0.0061	G	C
4	71,774,347	*MOB1B*	rs13137667	2.37E-08	0.0765	0.0137	T	C
5	1,285,974	*TERT*	rs7705526	4.82E-45	0.0820	0.0058	C	A
6	31,587,561	*PRRC2A*	rs2736176	3.41E-10	0.0345	0.0055	G	C
7	124,554,267	*POT1*	rs59294613	1.12E-13	−0.0407	0.0055	C	A
10	103,916,188	OBFC1	rs9419958	1.56E-19	0.0523	0.0058	C	T
11	108,105,593	*ATM*	rs228595	1.39E-08	−0.0285	0.0050	G	A
14	73,404,752	*DCAF4*	rs2302588	1.64E-08	0.0476	0.0084	G	C
16	69,406,986	*TERF2*	rs3785074	4.50E-10	0.0351	0.0056	A	G
16	82,199,980	*MPHOSPH6*	rs7194734	6.72E-10	−0.0369	0.0060	C	T
16	74,680,074	*RFWD3*	rs62053580	3.96E-08	−0.0389	0.0071	A	G
19	22,032,639	ZNF208	rs8105767	7.01E-21	−0.0420	0.0045	G	A
20	62,269,750	*STMN3*	rs75691080	5.75E-14	−0.0671	0.0089	C	T
20	62,291,599	*RTEL1*	rs34978822	7.04E-10	−0.1397	0.0227	C	G

Abbreviation: CHR, chromosome; POS, position; SNPS, single-nucleotide polymorphisms; PVAL, p-value; SE, standard error; REF, reference allele; EFT, effect allele.

Table 1.
Summary statistics of genetic instruments for telomere length.

3. Estimating telomere length from sequencing reads

3.1 Telomeres in the era of next-generation sequencing

Telomeres are repetitive sequences of $(TTAGGG)_n$ on chromosome ends. In humans, the length of telomeres can vary from 10 to 15 kilobases [19]. With advancements in genome sequencing technology, it is now possible to measure telomere length by applying computational algorithms to sequencing reads, as sequence data from these reads contains information about telomeres, just as they do for other regions of the genome.

DNA sequencing is the process of determining the order of nucleotide bases (T, C, A, G). To sequence a genome, the DNA is typically broken down into small fragments and sequenced in parallel [20]. The resulting sequence reads are then assembled to reconstruct the genome. High-throughput next-generation sequencing (NGS) allows for rapid and scalable sequencing of millions of DNA fragments. NGS can be used to sequence specific regions of the genome, such as protein-coding regions, or the entire genome to identify DNA sequences and somatic mutations. It can also be applied to RNA to measure gene expression.

Whole-genome sequencing (WGS) is a type of next-generation sequencing (NGS) that provides a comprehensive view of the entire genome, including the ends of chromosomes where telomeres are located. Several large-scale sequencing initiatives have been undertaken. These include the Trans-Omics for Precision Medicine (TOPMed) program, which includes over 130,000 whole-genome sequences from more than 80 studies [21], and the Pan-Cancer Analysis of Whole Genomes (PCAWG) project, an international effort to analyze over 2600 cancer whole genomes from the International Cancer Genome Consortium [22]. These projects have generated extensive WGS data integrated with clinical health outcomes on an individual level. Although telomere length was not the primary focus of these initiatives, the wealth of biomarker and health outcome data available makes these studies valuable resources for investigating the potential health impacts of telomere length variation on a population scale.

3.2 Computational tools for estimating telomere length and content

Numerous bioinformatic tools are currently available for estimating telomere length from WGS data. In 2010, Castle et al. first described their approach for estimating telomere content as a proxy for telomere length by counting sequencing reads that contained the repetitive $(TTAGGG)_4$ motif [23]. Since then, several algorithms have been developed and extended to determine telomere content, including Motif_counter [24], TelSeq [25], Computel [26], Telomerecat [27], TelomereHunter [28], Telogator [29], Qmotif [30]. Each tool is based on different algorithms and utilizes different methods to estimate telomere length/content from WGS data. In this section, we will introduce each tool and discuss their validation and application.

Motif_counter (https://sourceforge.net/projects/motifcounter) is a bash script that quantifies telomere content by counting sequencing reads that contain telomere motif. It takes Binary Alignment/Map (BAM) formatted files as input and thus requires sequence reads to be aligned to a reference genome prior to running. The motif to be searched for, i.e., the character string of the telomere repetitive region, and the threshold for classifying a read as a telomeric read must be specified. Despite its simplicity and user-friendly command interface, this tool is unable to control for variations in genome coverage or sequence depth, so additional normalization is needed. The telomere sequence read counts estimated by motif_counter correlated well with telomere length measured by TRF (Pearson's r = 0.855), after being normalized to genome coverage [31].

TelSeq (https://github.com/zd1/telseq) is a C++ software that takes BAM files as input and is specifically designed for estimating telomere length from WGS and could be extended to whole-exome sequencing (WES) data. It searches for telomeric reads of TTAGGG repeats and calculates the mean telomere length from telomeric content. One advantage of TelSeq is that it controls for sequencing biases by normalizing telomere counts based on the percentage of total reads that have a GC-content similar to telomere sequences (48–52%). This is important because a high GC value favors more amplification during PCR, potentially leading to biased estimates. TelSeq has been validated using 260 leukocyte samples from the TwinsUK cohort, where its estimated mean telomere length was compared to TRF estimates [25]. Although TelSeq estimates were consistently shorter than TRF estimates (mean 5.63 kb compared to 6.97 kb), their correlation remained stable across a range of pre-defined numbers of telomeric repeats (Spearman's ρ = 0.6). Additionally, TelSeq and TRF estimates had a correlation of 0.78 on exome data, providing promising evidence for its extended application. TelSeq is the most widely used tool and have been applied in several large WGS datasets [32–34].

Computel (https://github.com/lilit-nersisyan/computel) is an R program that operates in the Linux environment and takes FASTQ files as input. It identifies telomeric reads by aligning raw sequencing reads to a specially designed telomeric reference, distinguishing it from previous methods that were based on pattern matching. Mean telomere length is then calculated based on the ratio of coverage at the telomeric reference and genomic reference, read length, telomeric pattern length, and the number of chromosomes in a haploid genome. It also allows for telomeric repeat variant analysis to estimate the relative abundance of canonical and variant telomeric repeat patterns. This is important because telomeres may not always contain canonical repeat patterns $(TTAGGG)_n$ if there are variants within telomeric regions. Variant analysis provides necessary information about the distribution of telomeric repeat variants in samples. Computel has been validated with simulated data, where strong and linear correlation was observed between actual and estimated telomere length, and the results suggested that Computel could outperform TelSeq but consistently generated lower telomere length estimates [26].

Telomerecat (https://github.com/cancerit/telomerecat) is written in Python and operates in both the Linux and MacOSX environments. It is specifically designed to operate independently of the number of telomeres present in a cell by normalizing telomeric content against subtelomeric regions instead of the entire genome, making it applicable to WGS data from cancer cells. Telomerecat takes BAM files as input and extracts read pairs with at least two instances of telomeric hexamer. The software then classifies read pairs based on their sequence composition and orientation. Telomere length is calculated using the ratio of complete to boundary read-pairs, along with the insert length distribution. Telomerecat has been validated in 260 adult females from the TwinsUK10K study, showing a significant correlation with TRF estimates (Spearman's ρ = 0.618) and TelSeq (ρ = 0.631) [27].

TelomereHunter (https://pypi.org/project/telomerehunter) is a Python software designed to estimate telomeric content from WGS data of matched tumor and normal tissue control pairs within the same individual. The program accepts BAM files as input, selects reads with high telomere repeats, and organizes them by mapping their position into categories such as intrachromosomal, subtelomeric, junction spanning, and intratelomeric reads. Telomere content can be derived from the intratelomeric reads. Telomere content can be calculated from intratelomeric reads. TelomereHunter has been validated by strong correlations between its estimated telomere content and telomere length measures from qPCR (Pearson's r = 0.94) and TRF (Pearson's r = 0.72) after GC correction [28].

Telogator (https://github.com/zstephens/telogator) is a Python software designed to estimate chromosome-specific telomere length from long reads. This recently developed tool is built on long-read telomere analysis and the newest human reference genome by the Telomere-to-Telomere (T2T) consortium [35]. Telogator takes long reads in FASTA or FASTQ format, performs alignment, and extracts reads mapped to subtelomeres and telomeres. It then identifies telomere regions, clusters reads by their telomere-subtelomere boundaries, and reports telomere length for specific chromosome arms. High correlation (Pearson's r = 0.91) has been reported between Telogator and TelomereHunter when comparing the averaged chromosome-specific telomere lengths against the telomere content from TelomereHunter [29].

Qmotif (https://github.com/AdamaJava/adamajava) is a Java-based software designed to estimate telomere content from WGS data in a fast and efficient manner. It takes BAM files as input and searches for user-defined motifs using a two-pass matching system. In the first stage, a quick string matching is performed to filter strings into the second stage for regular expression (regex) matching. For telomere quantification, the first stage matches 3 consecutive repeats of the canonical telomere motif (TTAGGG), while the second stage uses regex to match any 2 adjacent repeats of the motif with variation allowed in the first 3 base pairs. The runtime can be further sped up by instructing the algorithm to search for telomere repeats in regions of the genome most likely to contain them. Qmotif has been validated by comparison with qPCR (Spearman's ρ = 0.69) and other computational tools such as TelSeq (Spearman's ρ = 0.99) and TelomereHunter (Spearman's ρ = 0.85), with a much faster runtime of under 1 minute on the same set of samples (compared to 1–7 hours for TelSeq and 4–19 hours for TelomereHunter) [30].

4. Sequencing based estimates for population genetic studies of telomere length

TelSeq has been used in several large WGS datasets, including the TOPMed study, which is one of the largest population-based studies with available WGS information. Telomere length was estimated using TelSeq in 109,122 TOPMed participants of diverse ancestry, including European, African, Asian, and Hispanic/Latino [33]. A strong correlation was observed between TelSeq estimates and TRF and flow-FISH in a subset of samples (r = 0.68 and 0.80, respectively). The results of GWAS on TelSeq-estimated telomere length were compared to those of GWAS on qPCR estimates and showed great consistency in effect estimates of the identified genetic predictors: Pearson's r = 0.92 for 37 overlapping variants with 23,096 Chinese Singaporeans and r = 0.86 for 43 overlapping variants with 78,592 Europeans. A genetic correlation of 0.81 between TelSeq and qPCR-estimated telomere length suggests that these methods share a high degree of genetic determinants.

In another study, TelSeq was applied to WES data from the UK Biobank, a population cohort of more than 500,000 UK adult residents [36]. Telomere length was estimated using TelSeq in 49,738 participants and compared to qPCR estimates from 472,594 participants within the UK Biobank and WGS estimates from 63,302 TOPMed participants using TelSeq. The WES-based telomere length (mean of 0.83 kb) was found to be shorter than the qPCR measures within the same population and even shorter than the WGS estimates in TOPMed (mean of 3.27 kb). Additionally, when estimating the effect sizes of SNPs in predicting telomere length, WES data showed the lowest correlation with SNPs and a deflation in effect size estimates. Since WES restricts sequencing to targeted regions of the genome, estimating telomere length from WES data may require further correction and adjustment.

5. Conclusions

In conclusion, genetic variants play an important role in determining leukocyte telomere length, enabling the estimation of an individual's telomere length through genotyping data. Furthermore, next-generation sequencing data offers an alternative to qPCR for measuring telomere length in large-scale population studies. The ability to estimate absolute telomere length in base pairs also facilitates the comparison of results across different studies. The use of sequencing-based telomere length estimates holds great promise for genetic studies, particularly given the likely overlap between sources of genetic variant and sequencing data.

Acknowledgements

This work was supported by National Cancer Institute grant U01ES029520.

Conflict of interest

The authors declare no conflict of interest.

Author details

Ting Zhai and Zachary D. Nagel*
John B. Little Center for Radiation Sciences, Department of Environmental Health, Harvard T.H. Chan School of Public Health, Boston, USA

*Address all correspondence to: znagel@hsph.harvard.edu

References

[1] Demanelis K, Jasmine F, Chen LS, Chernoff M, Tong L, Delgado D, et al. Determinants of telomere length across human tissues. Science. 2020;**369**(6509):eaaz6876

[2] Andreu-Sanchez S, Aubert G, Ripoll-Cladellas A, Henkelman S, Zhernakova DV, Sinha T, et al. Genetic, parental and lifestyle factors influence telomere length. Communications Biology. 2022;**5**(1):565

[3] Haycock PC, Heydon EE, Kaptoge S, Butterworth AS, Thompson A, Willeit P. Leucocyte telomere length and risk of cardiovascular disease: Systematic review and meta-analysis. BMJ. 2014;**349**:g4227

[4] Telomeres Mendelian Randomization C, Haycock PC, Burgess S, Nounu A, Zheng J, Okoli GN, et al. Association between telomere length and risk of cancer and non-neoplastic Diseases: A Mendelian randomization study. JAMA Oncology. 2017;**3**(5):636-651

[5] Kimura M, Stone RC, Hunt SC, Skurnick J, Lu X, Cao X, et al. Measurement of telomere length by the southern blot analysis of terminal restriction fragment lengths. Nature Protocols. 2010;**5**(9):1596-1607

[6] Baerlocher GM, Vulto I, de Jong G, Lansdorp PM. Flow cytometry and FISH to measure the average length of telomeres (flow FISH). Nature Protocols. 2006;**1**(5):2365-2376

[7] Cawthon RM. Telomere length measurement by a novel monochrome multiplex quantitative PCR method. Nucleic Acids Research. 2009;**37**(3):e21

[8] Telomere Research Network. Available from: https://trn.tulane.edu/

[9] Holohan B, Wright WE, Shay JW. Cell biology of disease: Telomeropathies: An emerging spectrum disorder. The Journal of Cell Biology. 2014;**205**(3):289-299

[10] Codd V, Wang Q, Allara E, Musicha C, Kaptoge S, Stoma S, et al. Polygenic basis and biomedical consequences of telomere length variation. Nature Genetics. 2021;**53**(10):1425-1433

[11] Chang X, Gurung RL, Wang L, Jin A, Li Z, Wang R, et al. Low frequency variants associated with leukocyte telomere length in the Singapore Chinese population. Communications Biology. 2021;**4**(1):519

[12] Davies NM, Holmes MV, Davey SG. Reading Mendelian randomisation studies: A guide, glossary, and checklist for clinicians. BMJ. 2018;**362**:k601

[13] de Leeuw C, Savage J, Bucur IG, Heskes T, Posthuma D. Understanding the assumptions underlying Mendelian randomization. European Journal of Human Genetics. 2022;**30**(6):653-660

[14] Codd V, Nelson CP, Albrecht E, Mangino M, Deelen J, Buxton JL, et al. Identification of seven loci affecting mean telomere length and their association with disease. Nature Genetics. 2013;**45**(4):422-427 7e1-2

[15] Gu J, Chen M, Shete S, Amos CI, Kamat A, Ye Y, et al. A genome-wide association study identifies a locus on chromosome 14q21 as a predictor of leukocyte telomere length and as a marker of susceptibility for bladder cancer. Cancer Prevention Research (Philadelphia, Pa.). 2011;**4**(4):514-521

[16] Lee JH, Cheng R, Honig LS, Feitosa M, Kammerer CM, Kang MS,

et al. Genome wide association and linkage analyses identified three loci-4q25, 17q23.2, and 10q11.21-associated with variation in leukocyte telomere length: The long life family study. Frontiers in Genetics. 2013;**4**:310

[17] Li C, Stoma S, Lotta LA, Warner S, Albrecht E, Allione A, et al. Genome-wide association analysis in humans links nucleotide metabolism to leukocyte telomere length. American Journal of Human Genetics. 2020;**106**(3):389-404

[18] Prescott J, Kraft P, Chasman DI, Savage SA, Mirabello L, Berndt SI, et al. Genome-wide association study of relative telomere length. PLoS One. 2011;**6**(5):e19635

[19] Srinivas N, Rachakonda S, Kumar R. Telomeres and telomere length: A general overview. Cancers (Basel). 2020;**12**(3):558

[20] van Dijk EL, Auger H, Jaszczyszyn Y, Thermes C. Ten years of next-generation sequencing technology. Trends in Genetics. 2014;**30**(9):418-426

[21] Taliun D, Harris DN, Kessler MD, Carlson J, Szpiech ZA, Torres R, et al. Sequencing of 53,831 diverse genomes from the NHLBI TOPMed program. Nature. 2021;**590**(7845):290-299

[22] Consortium ITP-CAoWG. Pan-cancer analysis of whole genomes. Nature. 2020;**578**(7793):82-93

[23] Castle JC, Biery M, Bouzek H, Xie T, Chen R, Misura K, et al. DNA copy number, including telomeres and mitochondria, assayed using next-generation sequencing. BMC Genomics. 2010;**11**:244

[24] Conomos D, Stutz MD, Hills M, Neumann AA, Bryan TM, Reddel RR, et al. Variant repeats are interspersed throughout the telomeres and recruit nuclear receptors in ALT cells. The Journal of Cell Biology. 2012;**199**(6):893-906

[25] Ding Z, Mangino M, Aviv A, Spector T, Durbin R, Consortium UK. Estimating telomere length from whole genome sequence data. Nucleic Acids Research. 2014;**42**(9):e75

[26] Nersisyan L, Arakelyan A. Computel: Computation of mean telomere length from whole-genome next-generation sequencing data. PLoS One. 2015;**10**(4):e0125201

[27] Farmery JHR, Smith ML, Diseases NB-R, Lynch AG. Telomerecat: A ploidy-agnostic method for estimating telomere length from whole genome sequencing data. Scientific Reports. 2018;**8**(1):1300

[28] Feuerbach L, Sieverling L, Deeg KI, Ginsbach P, Hutter B, Buchhalter I, et al. TelomereHunter - in silico estimation of telomere content and composition from cancer genomes. BMC Bioinformatics. 2019;**20**(1):272

[29] Stephens Z, Ferrer A, Boardman L, Iyer RK, Kocher JA. Telogator: A method for reporting chromosome-specific telomere lengths from long reads. Bioinformatics. 2022;**38**(7):1788-1793

[30] Holmes O, Nones K, Tang YH, Loffler KA, Lee M, Patch AM, et al. Qmotif: Determination of telomere content from whole-genome sequence data. Bioinformatics Advances. 2022;**2**(1):vbac005

[31] Lee M, Hills M, Conomos D, Stutz MD, Dagg RA, Lau LM, et al. Telomere extension by telomerase and ALT generates variant repeats by mechanistically distinct processes. Nucleic Acids Research. 2014;**42**(3):1733-1746

[32] Zheng S, Cherniack AD, Dewal N, Moffitt RA, Danilova L, Murray BA, et al. Comprehensive pan-genomic characterization of adrenocortical carcinoma. Cancer Cell. 2016;**29**(5):723-736

[33] Taub MA, Conomos MP, Keener R, Iyer KR, Weinstock JS, Yanek LR, et al. Genetic determinants of telomere length from 109,122 ancestrally diverse whole-genome sequences in TOPMed. Cell Genomics. 2022;**2**(1):100084

[34] Barthel FP, Wei W, Tang M, Martinez-Ledesma E, Hu X, Amin SB, et al. Systematic analysis of telomere length and somatic alterations in 31 cancer types. Nature Genetics. 2017;**49**(3):349-357

[35] Nurk S, Koren S, Rhie A, Rautiainen M, Bzikadze AV, Mikheenko A, et al. The complete sequence of a human genome. Science. 2022;**376**(6588):44-53

[36] Nakao T, Bick AG, Taub MA, Zekavat SM, Uddin MM, Niroula A, et al. Mendelian randomization supports bidirectional causality between telomere length and clonal hematopoiesis of indeterminate potential. Science Advances. 2022;**8**(14):eabl6579